42

Evans Attainment Tests 2

Anthony Wing and Gail West

Evans Brothers Limited London

Published by Evans Brothers Limited
Montague House, Russell Square, London WC1B 5BX

First published 1973

ISBN 0 237 28437 5 PRA 3415

Filmset and Printed in Great Britain in 11 pt Imprint by
Cox & Wyman Ltd, London, Fakenham and Reading

Acknowledgements

For permission to reproduce copyright material in this book, the authors and publishers are indebted to: Collins Publishers and David Higham Associates Ltd for the passage from *The Sword in the Stone* by T. H. White on p. 57; Ernest Benn Limited and John Farquharson Ltd for the passage from *The Railway Children* by E. Nesbit on p. 63; Heinemann Educational Books Ltd and John Farquharson Ltd for the passage from *The Treasure Seekers* by E. Nesbit on p. 75; Methuen & Co Ltd and The Bodleian Library, Oxford, for the passage from *The Wind in the Willows* by Kenneth Grahame on p. 81; J. M. Dent & Sons Ltd and the Trustees for the Copyrights of the late Dylan Thomas for the passage from *A Prospect of the Sea* by Dylan Thomas on p. 87.

Artwork and illustrations by Anna Potworowska.

Introduction

Our main aim in writing these books has been to provide teachers and concerned parents with a guide to the relative strengths and weaknesses of their children during the final year at junior school.

It has long been felt that the removal of the Eleven-Plus Examination has left many parents and teachers unsure about the standards of English and mathematics expected of children about to enter secondary school. It is to be hoped that these tests will provide a useful standard against which the achievements of junior school leavers in the basic skills of mathematics, reading and English usage may be measured.

In selecting the material for these tests we have tried to make the experience of actually doing a test as enlightening as possible for the child. It is our hope that most children will have gained something of value from simply thinking about the questions we have asked.

Notes of guidance to those who will be assessing the answers to the Mathematics Tests follow; notes on the English Usage Tests appear on p. 52.

A Guide to the Mathematics Tests

The Mathematics Tests are designed to assess children's achievements in five basic areas: computation; number; simple mathematical reasoning; space; and measurement. It is hoped that the tests will enable parents and teachers not only to get some idea of a child's overall grasp of mathematics, but also to search out those aspects of the subject which are most in need of attention.

The two books in this series together provide twelve tests which are approximately equal in difficulty and it is thus possible to give a series of monthly tests from the results of which progress may be measured during the final year at junior school.

The tests have deliberately been made difficult in order to be useful to as wide a range of children as possible. For a child to sail through a test with ease may be very gratifying to those concerned, but would do nothing to show up any relatively weak areas. Much of the value of these tests lies in the opportunity they provide for parents and teachers to discover weaknesses through discussion of incorrect answers and examination of questions not answered at all. Only towards the end of the final junior school year would one expect children to get a high proportion of the answers correct.

The total time allowed for a complete Mathematics Test should be eighty minutes. One obviously cannot expect an eleven-year-old to give of his or her best for this length of time, and it is therefore recommended that each test be divided into two sittings. In this case, thirty minutes should be allowed for the first half of each test (i.e. four pages), and fifty minutes for the second half.

Mathematics Test 1

1—Basic Computation

For each expression on the left, find the one on the right which gives the same answer. Write down its letter.

1. 6×8
2. 73
3. $7 \div 9$
4. $(3 \times 4) + 6$
5. $9 \times (8 - 3)$

(a) $(3 \times 25) - 2$
(b) 7×3
(c) $(10 - 1) \times (24 \div 12)$
(d) 68
(e) $(32 \times 3) \div 2$
(f) $21 \div 27$
(g) $10 + 10 + 10 + (5 \times 3)$
(h) $(3 + 4) \times 6$

Write down the number which should appear in each box below.

6. $4 \times 9 = 12 \times \square$
7. $6 \times 7 = \square \times 3$
8. $8 \times \square = 96 \div 3$
9. $9 \times 6 = \square \times 2 \times \square$
10. $\frac{2}{5} + \frac{3}{10} = 1 - \square$
11. £2.23 + 46p + 28p = $\square$
12. £1.32 = $\square \times$ 12p
13. £6.12 $\div$ 9 = $\square$
14. $\frac{3}{5}$ of £1.05 = $\square$
15. $\frac{2}{3}$ of 72p = $\square$

Mathematics Test 1

2—Understanding of Number

Numbers can be given shapes by using dots like this: ⁝˙˙ = ⁝·⁝ = 5

For each number shape on the left find the one on the right which stands for the same number. Write down its letter.

1. (a) (e)
2. (b) (f)
3. (c) (g)
4. (d) (h)
5.

Write down the *smallest* number in each line below.

6. 5586 6585 8655 5685 5568
7. $\frac{2}{3}$ $\frac{17}{24}$ $\frac{5}{6}$ $\frac{4}{9}$ $\frac{11}{12}$
8. $\frac{3}{8}$ 0.625 $\frac{1}{4}$ $7 \div 16$ $3 \div 16$

Which are the correct answers to these questions, (a), (b), (c) or (d)?

9. 12−17 (a) 5 (b) 29 (c) −5 (d) 0
10. 29−31 (a) 60 (b) −2 (c) 0 (d) 2

11. **Write in figures the number which has:** eight tens, seven units, three thousands and two hundreds.

12. **Write down the *smallest* number which can be made from the figures:** three, eight, two and two.

Mathematics Test 1

3—Understanding of Number

Three squares can be joined together to make a *row* like this:

Three rows can be joined together to make a *large square* like this:

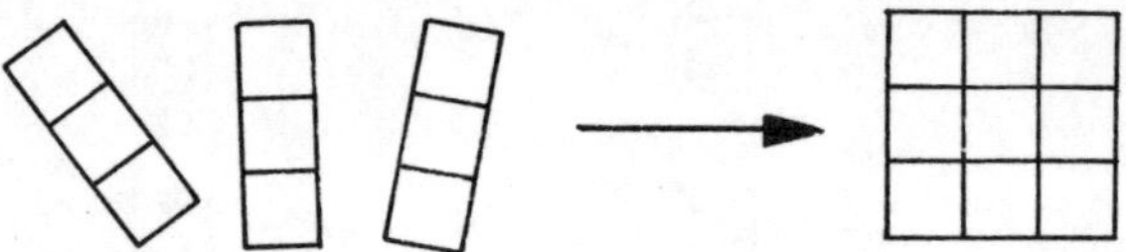

13 squares will make:

Large Squares	Rows	Left Over
1	1	1

How many large squares and rows will the following squares make, and how many squares will be left over? Write your answers in your book.

	Squares	Large Squares	Rows	Left Over
13.	14			
14.	22			
15.	18			
16.	10			
17.	26			

18. If *four* small squares are used to make a row, how many small squares will be needed to make a large square?

Mathematics Test 1

4—Sets, Sequences and Reasoning

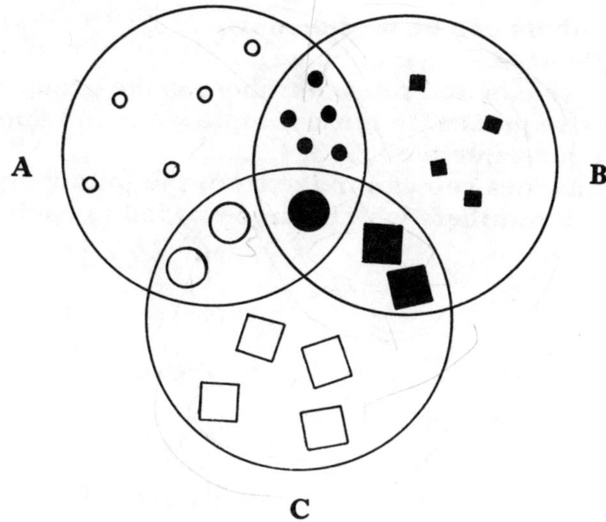

Write down the answers to the following questions:

1. Circle A contains all the shapes that are: (a) white
(b) square
(c) small
(d) round
(e) black

2. Circle B contains all the shapes that are: (a) round
(b) large
(c) black
(d) white
(e) small

3. How many squares that are in B are also in C?

4. How many large round shapes are there in A?

5. When all the shapes that are in B and C have been taken away, the shapes that are left are all: (a) large and round
(b) large and white
(c) small and black
(d) small and round

Mathematics Test 1

5—Sets, Sequences and Reasoning

Some groups of numbers can be written down *in order* so that they form a *sequence*. For example, [1, 3, 4, 2] can be written as [1, 2, 3, 4]. When we have formed a sequence we can add more numbers to the group by working out the next numbers in the sequence. In our example we could add 5 and 6 because these come next in the sequence 1, 2, 3, 4,.....
Write down the sequences you can make from the following groups, and then write down two more numbers which can be added to each group.

6. [31, 35, 27, 23]
7. [45, 37, 29, 53]
8. [37, 25, 49, 61]
9. [18, 6, 2, 54]
10. [12, 24, 6, 3]
11. [10, 8, 13, 7]
12. [7, 11, 5, 17]
13. [$\frac{1}{9}$, $\frac{1}{1}$, $\frac{1}{27}$, $\frac{1}{3}$]
14. [$\frac{2}{2}$, $\frac{2}{1}$, $\frac{2}{8}$, $\frac{2}{4}$]
15. [$\frac{10}{12}$, $\frac{2}{4}$, $\frac{6}{8}$, $\frac{14}{16}$]

Here are some statements about a number:
(i) ☐ is a whole number
(ii) ☐ is an odd number
(iii) ☐ is less than ten
(iv) ☐ is greater than seven.

If the number *nine* is put in each box then *all* the statements are *true*.

Write down the number which will make *all* the statements true in each of the following questions.

16. (i) ☐ is not an even number

    (ii) ☐ is less than ten

    (iii) ☐ is not less than seven

    (iv) ☐ is not a multiple of three.

17. (i) ☐ is a prime number

    (ii) ☐ is less than seven

    (iii) ☐ is not less than three

    (iv) ☐ is not a factor of twelve.

Mathematics Test 1

6—Spatial Discrimination

Which of the following groups of plane shapes could be put together to make solid shapes?

1.

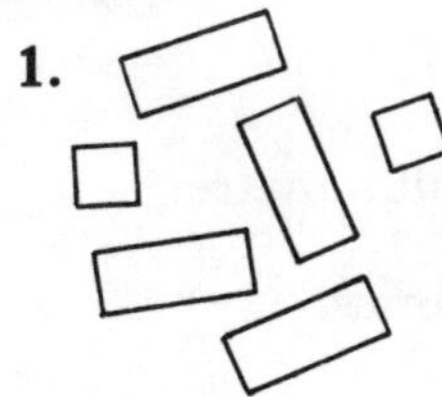

2.

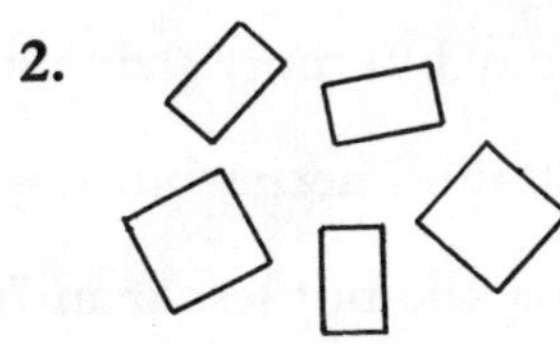

3.

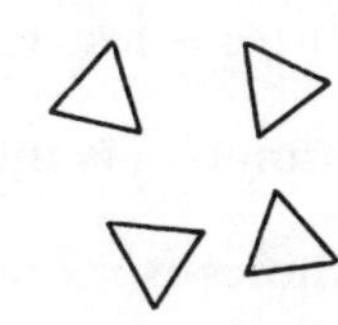

4.

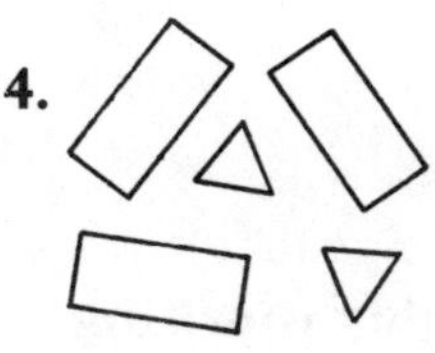

5.

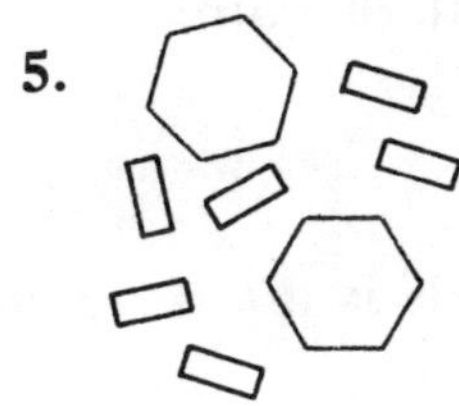

6.

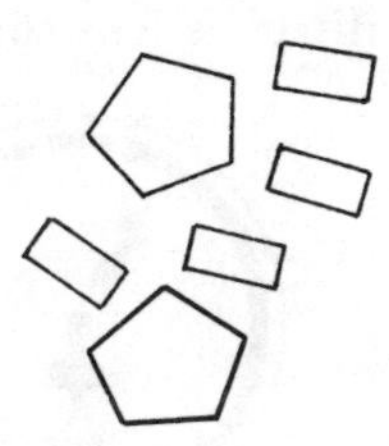

Use *one* or *two* of the following words to describe each of the solids you could make in questions 1 to 6:

triangular, square, rectangular, pentagonal, hexagonal, prism, pyramid, octahedron, tetrahedron, cube, cuboid.

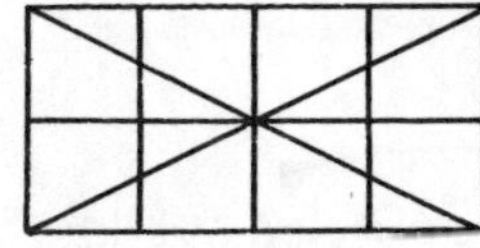

Which of the following shapes are to be found in the figure on the left?

7.

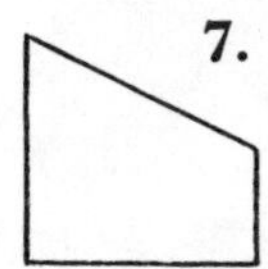

8.

9.

10.

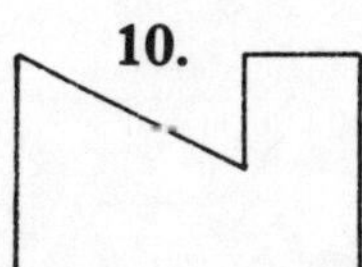

11.

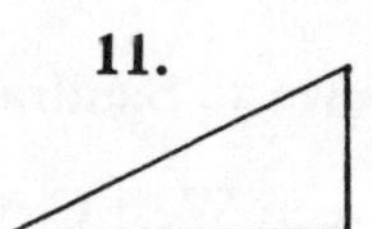

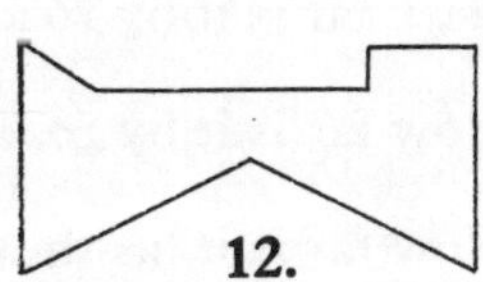

12.

Mathematics Test 1

7—Measurements and Measuring

Write down one number which would make each of the following statements true. There may be more than one right answer to each question.

1. □ minutes is less than half a day but more than ten hours.
2. □ centimetres is more than 80 millimetres but less than half a metre.
3. □ grammes is more than half a kilo but less than 700 grammes.
4. □ millimetres is more than two metres.
5. □ millilitres is less than an eighth of a litre.

This is the time on a Monday morning.

6. In how many hours and minutes will it be 3.00 p.m. on Tuesday?
7. How many hours and minutes have passed since 2.30 p.m. Sunday?

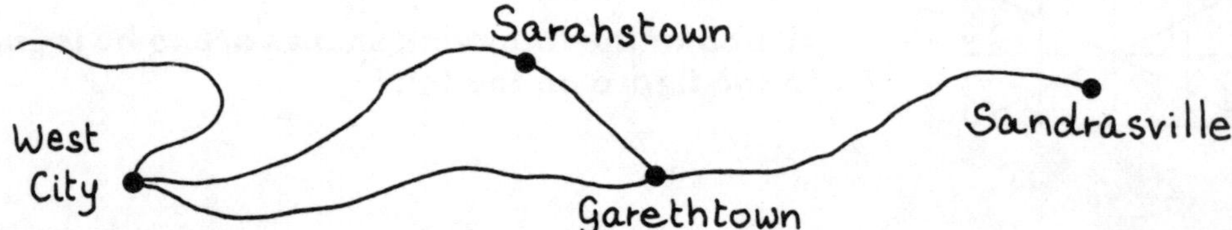

On this map 1 centimetre stands for 10 kilometres.

8. How far is it by road from West City to Sarahstown?
9. How far is it by road from Garethtown to Sandrasville?
10. How far is it 'as the crow flies' between West City and Sandrasville?

Mathematics Test 1

8—Measurements and Measuring

Sarah decided to measure Gareth's heartbeat rate when he was doing some running. Shown below is a graph she drew of her results. Look carefully at the graph and then answer the questions below.

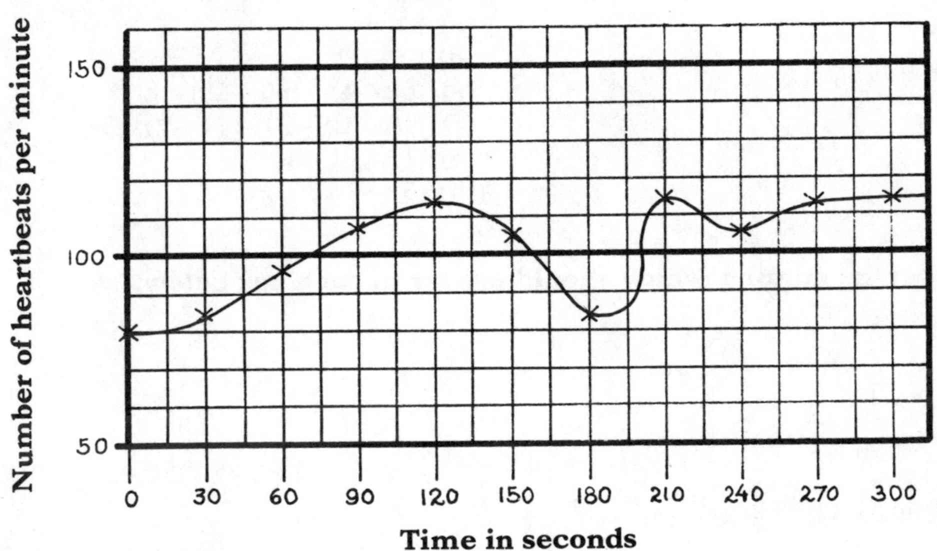

11. What was Gareth's heart rate after 2 minutes?

12. After how many seconds was Gareth's heart beating fastest?

13. How many seconds did it take for Gareth's heart rate to increase from 80 to 100 heartbeats per minute?

14. How many seconds did it take for Gareth's heart rate to decrease from 110 to 90 heartbeats per minute?

15. For how many seconds was Gareth's heart beating faster than 100 heartbeats per minute?

16. How many times did Gareth's heart actually beat during the last 45 seconds shown on the graph?

Mathematics Test 2

1—Basic Computation

For each expression on the left find the one on the right which gives the same answer. Write down its letter.

1. $68 \div (17-15)$
2. 9×6
3. $(5 \times 4)+17$
4. 28
5. $7 \div 4$

(a) $(9 \times 12) \div 2$
(b) 96
(c) 3×4
(d) $21 \div 12$
(e) $(60-4) \div (60-58)$
(f) $(10+10+20)-(1+2)$
(g) $(12 \times 3)-2$
(h) $(5+4) \times 17$

Write down the number which should appear in each box below.

6. $8 \times 9 = 12 \times \square$
7. $6 \times 7 = \square \times 3$
8. $4 \times \square = 96 \div 3$
9. $18 \times 5 = \square \times \square \times 2$
10. $\frac{3}{4} - \frac{3}{8} = \frac{3}{2} \times \square$
11. £3.40+£2.60+32p $= \square$
12. £1.85 $= \square \times$ 37p
13. £2.65 $\div 5 = \square$
14. $\frac{3}{7}$ of £1.68 $= \square$
15. $\frac{5}{6}$ of 78p $= \square$

Mathematics Test 2

2—Understanding of Number

Numbers can be given shapes by using dots like this: ⁘ = ⁘ = 4

For each number shape on the left find the one on the right which stands for the same number. Write down its letter.

1. ⁘	(a)	(e)	
2. ⁘	(b)	(f)	
3. ⁘	(c)	(g)	
4. ⁘	(d)	(h)	
5. ⁘			

Write down the *smallest* number in each line below.

6. 5382	5283	3852	3528	3825
7. $\frac{3}{4}$	$\frac{5}{12}$	$\frac{7}{16}$	$\frac{13}{32}$	$\frac{5}{8}$
8. $\frac{2}{5}$	0.3	$7 \div 20$	$\frac{7}{15}$	$31 \div 100$

Which are the correct answers to these questions, (a), (b), (c) or (d)?

9. $23-32$	(a) 55	(b) 9	(c) 0	(d) -9
10. $6-11$	(a) 0	(b) -5	(c) 5	(d) 17

11. **Write in figures the number which has:** seven thousands, three hundreds, two units and five tens.

12. **Write down the *smallest* number which can be made from the figures:** two, eight, seven and one.

Mathematics Test 2

3—Understanding of Number

If you are in any doubt about what you are being asked to do in the following questions, you should turn to page 6 in this book where problems *like* the ones below are explained fully.

Four **small squares can be joined together to make a** *row.*

Four **rows can be joined together to make a** *large square.*

21 small squares will make:

Large Squares	Rows	Left Over
1	1	1

How many large squares and rows will the following squares make, and how many squares will be left over? Write your answers in your book.

	Squares	Large Squares	Rows	Left Over
13.	43			
14.	25			
15.	19			
16.	16			
17.	34			

18. If *seven* small squares are used to make a row, how many small squares are needed to make a large square?

19. If 64 small squares are used to make a large square, how many small squares are in a row?

Mathematics Test 2

4—Sets, Sequences and Reasoning

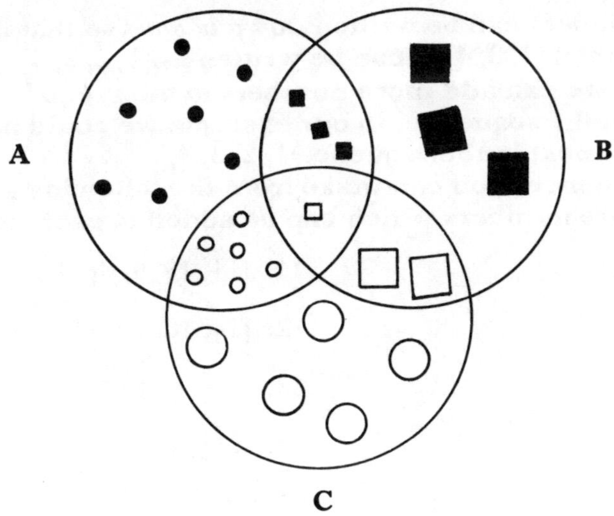

Write down the answers to the following questions:

1. Circle A contains all the shapes that are: (a) large
(b) black
(c) square
(d) round
(e) small

2. Circle B contains all the shapes that are: (a) black
(b) square
(c) white
(d) large
(e) small

3. How many square shapes that are in A are also in B?

4. How many small white shapes are there in C?

5. When all the shapes that are in B and C have been taken away the shapes that are left are all: (a) white and round
(b) small and square
(c) small and round
(d) large and square

Mathematics Test 2

5—Sets, Sequences and Reasoning

Some groups of numbers can be written down *in order* so that they form a *sequence*. For example, [1, 3, 4, 2] can be written as [1, 2, 3, 4]. When we have formed a sequence we can add more numbers to the group by working out the next numbers in the sequence. In our example we could add 5 and 6 because these come next in the sequence 1, 2, 3, 4, ...
Write down the sequences you can make from the following groups, and then write down two more numbers which can be added to each group.

6. [53, 49, 55, 51]
7. [42, 35, 28, 49]
8. [23, 49, 36, 10]
9. [20, 10, 5, 40]
10. [63, 189, 21, 7]
11. [11, 15, 9, 8]
12. [14, 10, 19, 11]
13. [$\frac{1}{1}$, $\frac{3}{2}$, $\frac{2}{1}$, $\frac{1}{2}$]
14. [$\frac{3}{8}$, $\frac{1}{2}$, $\frac{1}{4}$, $\frac{1}{8}$]
15. [$\frac{1}{12}$, $\frac{1}{6}$, $\frac{1}{24}$, $\frac{1}{3}$]

Here are some statements about a number:
(i) ☐ is a whole number
(ii) ☐ is an odd number
(iii) ☐ is less than ten
(iv) ☐ is greater than seven.

If the number *nine* is put in each box then *all* the statements become *true*.

Write down the number which will make *all* the statements true in each of the following questions.

16. (i) ☐ is not less than twenty
    (ii) ☐ is not an even number
    (iii) ☐ is less than twenty-five
    (iv) ☐ is not a prime number.

17. (i) ☐ is greater than thirteen
    (ii) ☐ is an odd number
    (iii) ☐ is not greater than eighteen
    (iv) ☐ is a prime number.

Mathematics Test 2

6—Spatial Discrimination

Which of the following groups of plane shapes could be put together to make solid shapes?

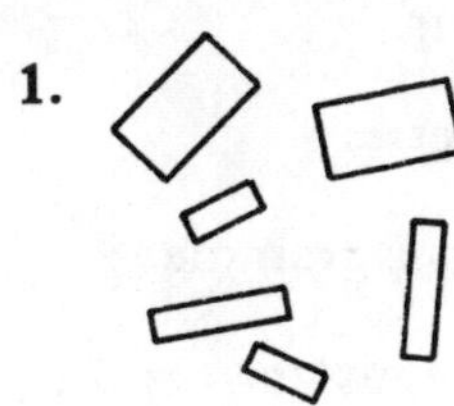

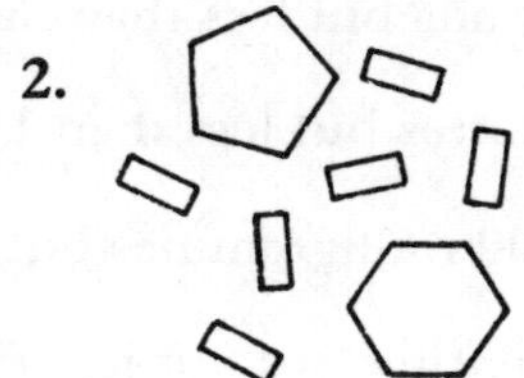

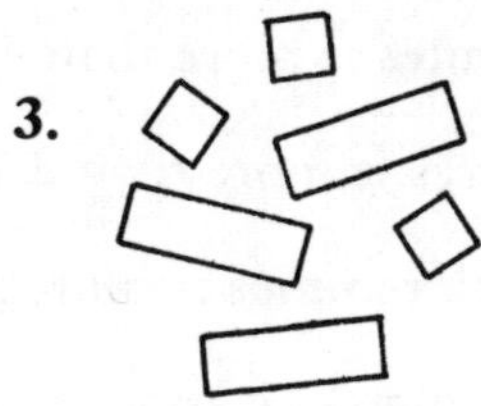

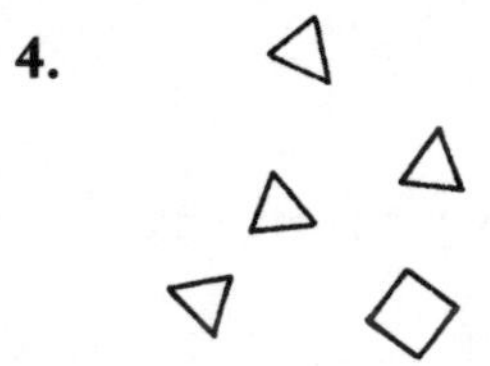

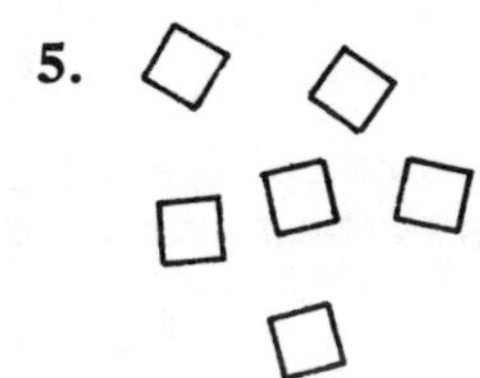

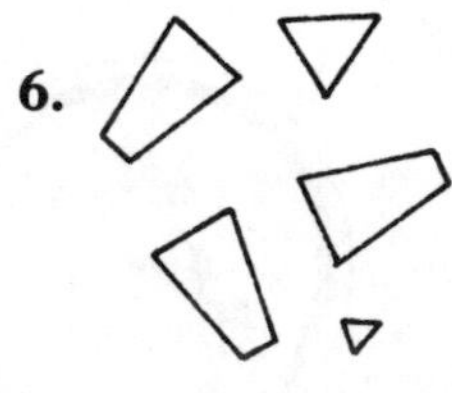

Use *one* or *two* of the following words to describe each of the solids you could make in questions 1 to 6:

triangular, square, rectangular, pentagonal, hexagonal, prism, pyramid, tetrahedron, cube, cuboid, truncated.

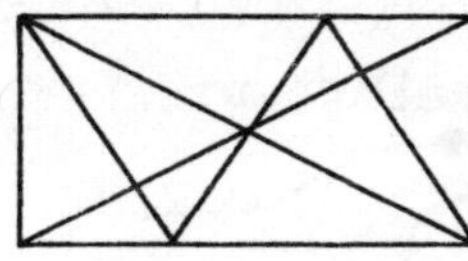

Which of the following shapes are to be found in the figure on the left?

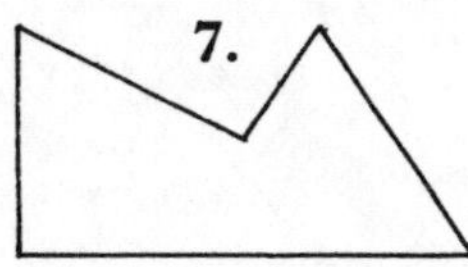

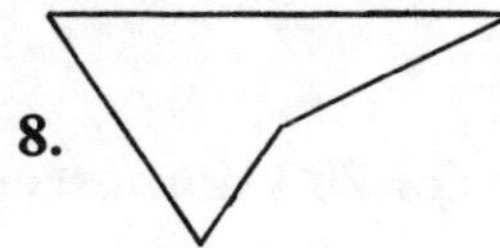

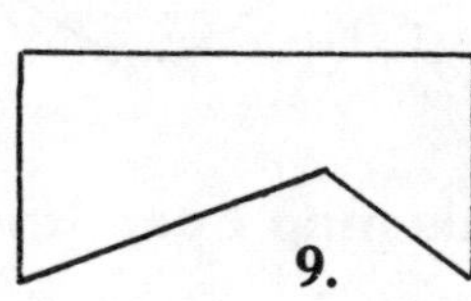

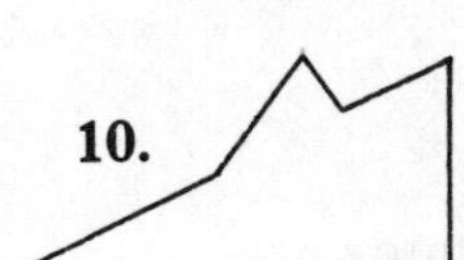

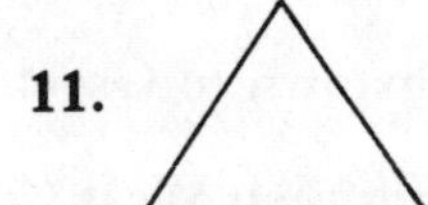

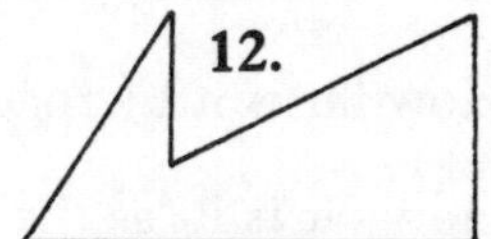

Mathematics Test 2

7—Measurements and Measuring

Write down one number which would make each of the following statements true. There may be more than one right answer to each question.

1. □ minutes is more than 400 seconds but less than half an hour.
2. □ metres is more than 3 Dekametres but less than 1 Hectametre.
3. □ centigrammes is more than 400 milligrammes but less than 1 gramme.
4. □ millimetres is more than 20 centimetres but less than half a metre.
5. □ cubic centimetres is less than a quarter of a cubic metre.

This is the time on a Saturday morning.

6. In how many hours and minutes will it be 4.00 p.m. on Sunday?
7. How many hours and minutes have passed since 3.30 a.m. on Friday?

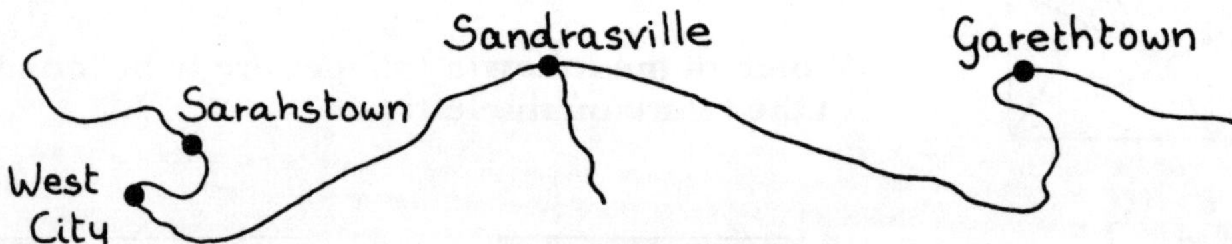

On this map 1 centimetre stands for 20 kilometres.

8. How far is it by road from West City to Sandrasville?
9. How far is it by road from Sarahstown to Garethtown?
10. How far is it 'as the crow flies' between West City and Garethtown?

Mathematics Test 2

8—Measurements and Measuring

Sandra decided to measure the speed of her bicycle as she went down to the shops. Shown below is a graph she drew from her results. Look carefully at the graph and then answer the questions below.

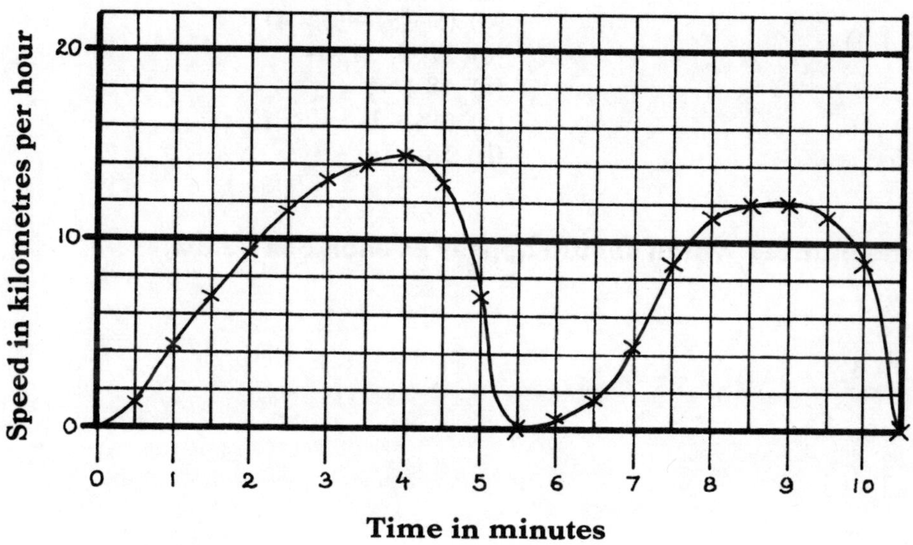

11. What was Sandra's speed after seven minutes?

12. After how many minutes did Sandra reach her top speed?

13. How many minutes and seconds did it take for Sandra's speed to increase from 8 to 14 kilometres per hour?

14. How many minutes and seconds did it take for Sandra's speed to decrease from 14 to 4 kilometres per hour?

15. For how many minutes and seconds was Sandra travelling faster than 10 kilometres per hour?

16. How many metres, approximately, did Sandra travel during the ninth minute?

Mathematics Test 3

1—Basic Computation

For each expression on the left find the one on the right which gives the same answer. Write down its letter.

1. 3×13
2. 72
3. $(5 \times 5) - (3 + 4)$
4. $8 \div 9$
5. $14 \div (35 \div 5)$

(a) 313
(b) $(5 - 4 + 3) - (96 \div 48)$
(c) $(3 \times 20) - (42 \div 2)$
(d) $(5 + 5) - (3 \times 4)$
(e) $2 \times 3 \times 3 \times 4$
(f) $48 \div 54$
(g) $(48 \div 3) + (2 \times 1)$
(h) $18 \div 19$

Write down the number which should appear in each box below.

6. $8 \times 11 = 4 \times \square$
7. $6 \times 9 = \square \times 3$
8. $9 \times \square = 126 \div 2$
9. $20 \times 6 = \square \times \square \times 5$
10. $\frac{1}{6} + \frac{5}{12} = 1 - \square$
11. £3.21 + 43p + £0.03 = $\square$
12. £2.90 = $\square \times$ 10p
13. £7.35 ÷ 7 = $\square$
14. $\frac{2}{7}$ of £1.12 = $\square$
15. $\frac{3}{5}$ of 30p = $\square$

Mathematics Test 3

2—Understanding of Number

Numbers can be given shapes by using dots like this:  =  = 3

For each number shape on the left find the one on the right which stands for the same number. Write down its letter.

1.  (a)  (e) 

2.  (b)  (f) 

3.  (c)  (g) 

4.  (d)  (h) 

5. 

Write down the *smallest* number in each line below.

6. 7768 7786 8767 7867 7678

7. $\frac{3}{5}$ $\frac{8}{15}$ $\frac{11}{20}$ $\frac{19}{30}$ $\frac{3}{10}$

8. $\frac{2}{7}$ $\frac{1}{8}$ 0.15 $2 \div 9$ $13 \div 63$

Which are the correct answers to these questions, (a), (b), (c) or (d)?

9. $3 - 18$ (a) 15 (b) 0 (c) -15 (d) 21

10. $15 - 21$ (a) 6 (b) -6 (c) -36 (d) 0

11. **Write in figures the number which has:** four thousands, seven units, five hundreds and two tens.

12. **Write down the *smallest* number which can be made from the figures:** five, eight, seven, two and three.

Mathematics Test 3

3—Understanding of Number

If you are in any doubt about what you are being asked to do in the following questions, you should turn to page 6 in this book where problems *like* the ones below are explained fully.

***Six* small squares can be joined together to make a *row*.**

***Six* rows can be joined together to make a *large square*.**

43 small squares will make:

Large Squares	Rows	Left Over
1	1	1

How many large squares and rows will the following squares make, and how many squares will be left over? Write your answers in your book.

	Squares	Large Squares	Rows	Left Over
13.	47			
14.	38			
15.	42			
16.	56			
17.	73			

18. If *five* small squares are used to make a row, how many small squares will be needed to make a large square?

19. If 100 small squares are used to make a large square, how many small squares are in a row?

Mathematics Test 3

4—Sets, Sequences and Reasoning

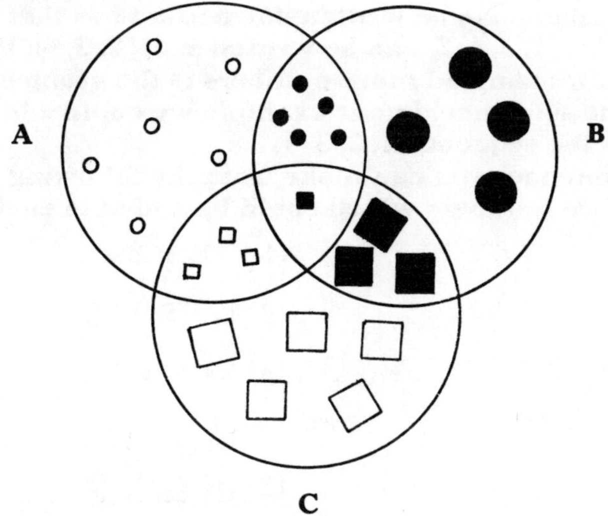

Write down the answers to the following questions:

1. Circle A contains all the shapes that are: (a) white
(b) black
(c) round
(d) small
(e) square

2. Circle B contains all the shapes that are: (a) round
(b) large
(c) square
(d) black
(e) small

3. How many small shapes that are in A are also in B?

4. How many small black shapes are there in C?

5. When all the shapes in A and C have been taken away the shapes that are left are all: (a) round and small
(b) large and round
(c) large and square
(d) black and small

Mathematics Test 3

5—Sets, Sequences and Reasoning

Some groups of numbers can be written down *in order* so that they form a *sequence*. For example, [1, 3, 4, 2] can be written as [1, 2, 3, 4]. When we have formed a sequence we can add more numbers to the group by working out the next numbers in the sequence. In our example we could add 5 and 6 because these come next in the sequence 1, 2, 3, 4, ...
Write down the sequences you can make from the following groups, and then write down two more numbers which could be added to each group.

6. [63, 57, 61, 59]
7. [76, 81, 71, 66]
8. [47, 63, 31, 79]
9. [32, 8, 128, 2]
10. [16, 32, 8, 4]
11. [11, 5, 2, 2]
12. [7, 15, 3, 1]
13. [$\frac{11}{4}$, $\frac{5}{2}$, 2, $\frac{9}{4}$]
14. [$\frac{5}{16}$, $\frac{7}{16}$, $\frac{3}{8}$, $\frac{1}{4}$]
15. [$\frac{17}{32}$, $\frac{19}{32}$, $\frac{1}{2}$, $\frac{9}{16}$]

Here are some statements about a number:

(i) ☐ is a whole number
(ii) ☐ is an odd number
(iii) ☐ is less than ten
(iv) ☐ is greater than seven.

If the number *nine* is put in each box then *all* the statements become *true*.

Write down the number which will make *all* the statements true in each of the following questions:

16. (i) ☐ has a factor of four

    (ii) ☐ is a multiple of two

    (iii) ☐ is not greater than ten

    (iv) ☐ is not less than five.

17. (i) ☐ is a factor of thirty

    (ii) ☐ has a factor of three

    (iii) ☐ is not less than ten

    (iv) ☐ is not greater than twenty.

Mathematics Test 3

6—Spatial Discrimination

Which of the following groups of plane shapes could be put together to make solid shapes?

1. 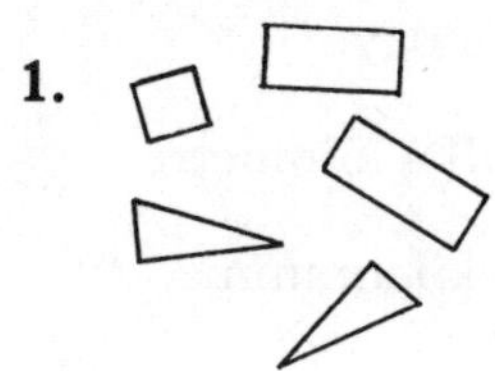

2. 

3. 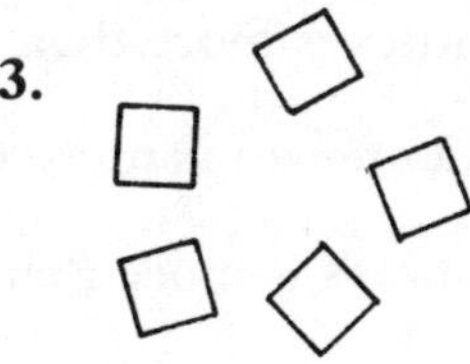

4. 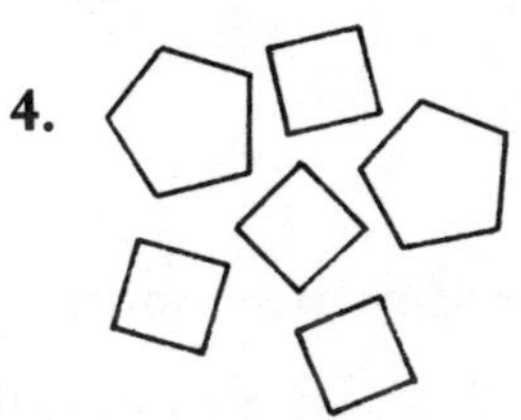

5. 

6. 

Use *one* or *two* of the following words to describe each of the solids you could make in questions 1 to 6:
triangular, square, rectangular, pentagonal, hexagonal, prism, pyramid, octahedron, cube, cuboid.

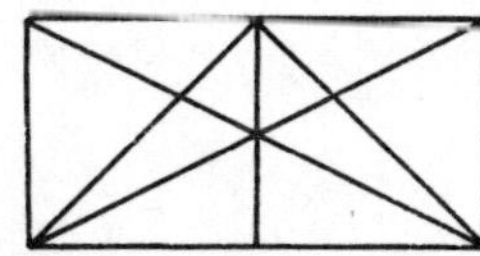

Which of the following shapes are to be found in the figure on the left?

7. 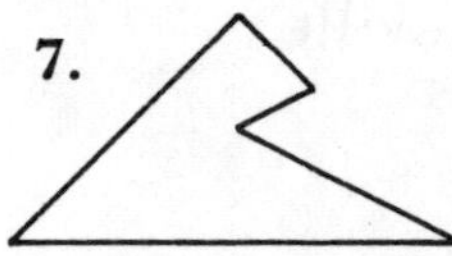

8. 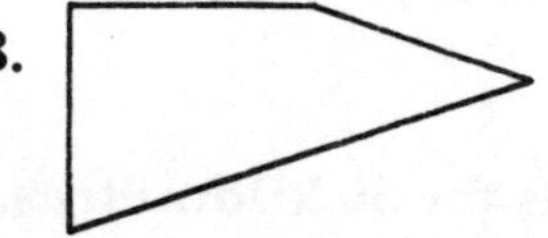

9. 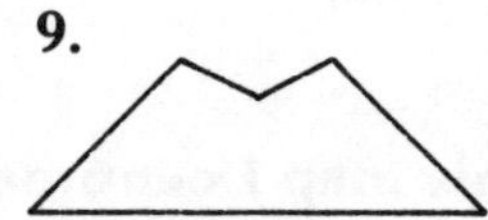

10. 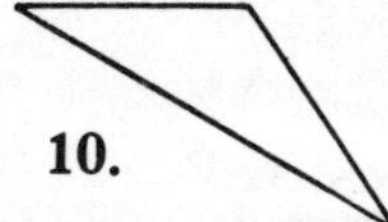

11. 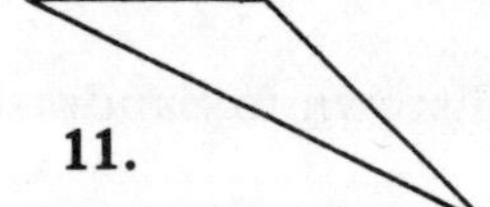

12. 

Mathematics Test 3

7—Measurements and Measuring

Write down one number which would make each of the following statements true. There may be more than one right answer to each question.

1. □ minutes is more than seven hours but less than a third of a day.
2. □ Hectametres is more than 30 Dekametres but less than half a kilometre.
3. □ grammes is more than 8 Hectagrammes but less than one kilogramme.
4. □ decimetres is more than 20 centimetres but less than half a metre.
5. □ cubic centimetres is less than a tenth of a cubic metre.

This is the time on a Wednesday afternoon.

6. In how many hours and minutes will it be 2.30 a.m. on Thursday?
7. How many hours and minutes have passed since 7.15 p.m. Monday?

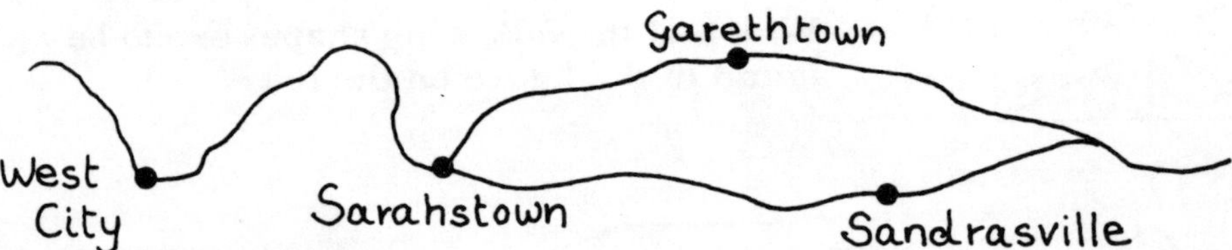

On this map 1 centimetre stands for 50 kilometres.

8. How far is it by road from West City to Garethtown?
9. How far is it by road from Sarahstown to Sandrasville?
10. How far is it 'as the crow flies' between West City and Sarahstown?

Mathematics Test 3

8—Measurements and Measuring

Gareth decided to measure Sarah's rate of walking as they walked to school one day. Shown below is a graph he drew from his results. Look carefully at the graph and then answer the questions below.

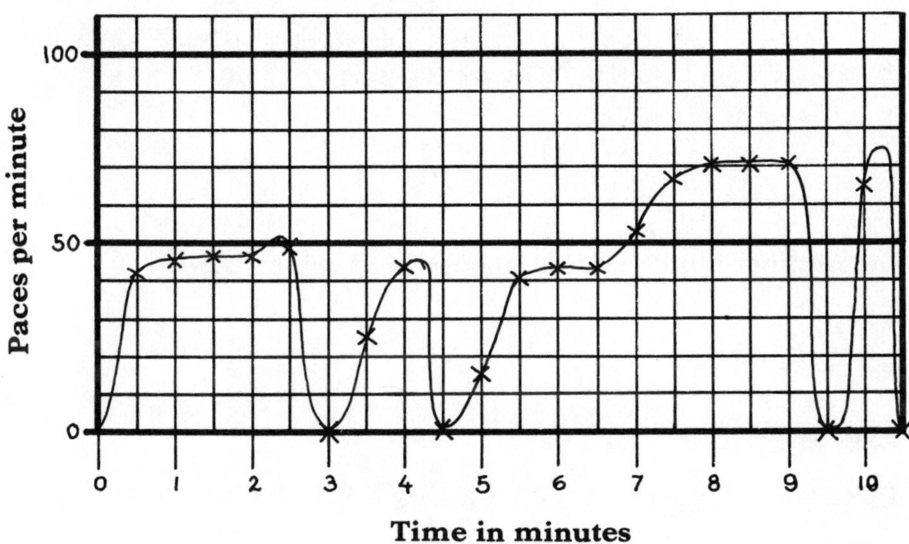

11. What was Sarah's rate of walking after seven minutes?

12. After how many minutes was Sarah walking fastest?

13. How many seconds did it take for Sarah to slow down from 40 paces per minute to make her first stop?

14. How many seconds did it take for Sarah to work up to 40 paces per minute after stopping for the second time?

15. For how many minutes was Sarah walking faster than 50 paces per minute?

16. At one point Sarah looked at her watch, realized she was going to be late, and speeded up. When do you think this was?

Mathematics Test 4

1—Basic Computation

For each expression on the left find the one on the right which gives the same answer. Write down its letter.

1. 85
2. 4×15
3. $72\div(36\div 3)$
4. $(4\times 8)+18$
5. $7\div 12$

(a) $450\div(3\times 3)$
(b) 415
(c) $84\div 144$
(d) $5\times 4\times 3\times 1$
(e) $3+2+1-(17\times 0)$
(f) 8×5
(g) $(4+8)\times 18$
(h) $100-15$

Write down the number which should appear in each box below.

6. $12\times 9 = 4\times\square$
7. $8\times 6 = \square\times 3$
8. $6\times\square = 126\div 3$
9. $10\times 7 = \square\times 5\times\square$
10. $\frac{3}{4}+\frac{1}{12}=1-\square$
11. £4.12+38p+£1.01 = $\square$
12. £2.24 = $\square\times$ 8p
13. £3.12 ÷ 4 = $\square$
14. $\frac{5}{8}$ of £1.28 = $\square$
15. $\frac{2}{9}$ of 81p = $\square$

Mathematics Test 4

2—Understanding of Number

Numbers can be given shapes by using dots like this:  =  = **3**

For each number shape on the left find the one on the right which stands for the same number. Write down its letter.

1.  (a)  (e) 

2.  (b)  (f) 

3.  (c)  (g) 

4.  (d)  (h) 

5. 

Write down the *smallest* number in each line below.

6. 6678 6687 6768 6876 6786

7. $\frac{5}{2}$ $\frac{7}{8}$ $\frac{17}{16}$ $\frac{8}{6}$ $\frac{13}{12}$

8. $\frac{7}{2}$ $2\frac{1}{2}$ 2.75 $23 \div 12$ $10 \div 3$

Which are the correct answers to these questions, (a), (b), (c) or (d)?

9. $11-22$ (a) 11 (b) -11 (c) 0 (d) 33

10. $6-19$ (a) 0 (b) 13 (c) -13 (d) -25

11. **Write in figures the number which has:** six tens, eight units and five thousands.

12. **Write down the *largest* number which can be made from the figures:** four, four, three and eight.

Mathematics Test 4

3—Understanding of Number

Three small cubes can be joined together to make a *row* like this:

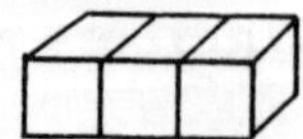

Three rows can be joined together to make a *square* like this:

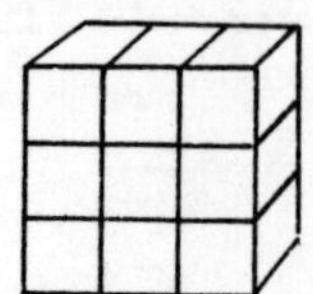

Three squares can be joined together to make a *large cube* like this:

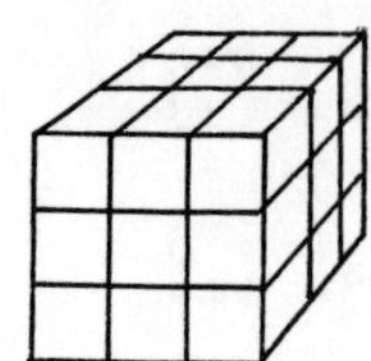

36 small cubes will make:

Large Cubes	Squares	Rows	Left Over
1	1	0	0

How many large cubes, squares and rows will the following cubes make, and how many cubes will be left over? Write your answers in your book.

	Small Cubes	Large Cubes	Squares	Rows	Left Over
13.	47				
14.	38				
15.	53				
16.	66				
17.	72				

18. If *four* small cubes are used to make a row, how many small cubes will be needed to make a large cube?

Mathematics Test 4

4—Sets, Sequences and Reasoning

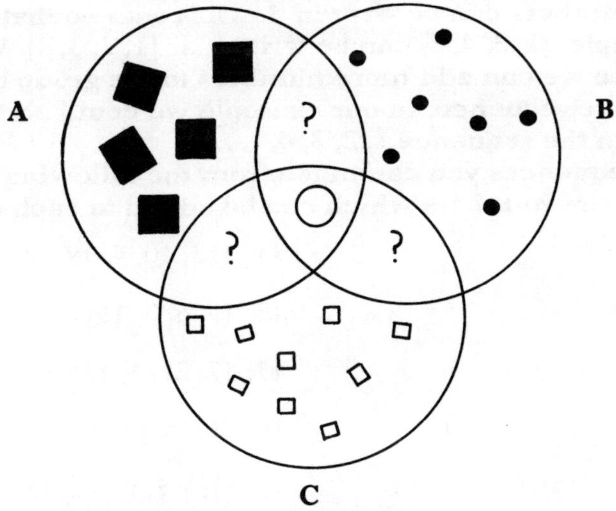

Write down the answers to the following questions:

1. Circle A contains all the shapes that are: (a) round
   (b) black
   (c) square
   (d) white
   (e) large

2. Circle B contains all the shapes that are: (a) black
   (b) round
   (c) small
   (d) square
   (e) white

3. Draw one shape which would be contained by both circle A and circle C, but not by circle B. State whether it is large or small.

4. Draw one shape which would be contained by both circle B and circle C, but not by circle A. State whether it is large or small.

5. When all the shapes that are in A and B have been taken away the shapes that are left are all: (a) large and square
   (b) white and small
   (c) small and round
   (d) large and white

Mathematics Test 4

5—Sets, Sequences and Reasoning

Some groups of numbers can be written down *in order* so that they form a *sequence*. For example, [1, 3, 4, 2] can be written as [1, 2, 3, 4]. When we have formed a sequence we can add more numbers to the group by working out the next numbers in the sequence. In our example we could add 5 and 6 because these come next in the sequence 1, 2, 3, 4, ...
Write down the sequences you can make from the following groups, and then write down two more numbers which can be added to each group.

6. [44, 38, 41, 47]
7. [63, 72, 54, 45]
8. [51, 36, 66, 21]
9. [28, 7, 14, 56]
10. [7, 9, 13, 3, 1]
11. [13, 10, 4, 19, 1]
12. [9, 5, 3, 15]
13. [7, 22, 4, 13]
14. [1, $1\frac{2}{3}$, $\frac{2}{3}$, $2\frac{2}{3}$]
15. [$\frac{1}{2}$, $\frac{9}{8}$, $\frac{3}{8}$, $\frac{3}{4}$]

Here are some statements about a number:
(i) ☐ is a whole number
(ii) ☐ is an odd number
(iii) ☐ is less than ten
(iv) ☐ is greater than seven.

If the number *nine* is put in each box then *all* the statements become *true*.

Write down the number which will make *all* the statements true in each of the following questions:

16. (i) ☐ is not an odd number
    (ii) ☐ is not greater than six
    (iii) ☐ is not less than five
    (iv) ☐ is a perfect number.
17. (i) ☐ has a factor of three
    (ii) ☐ is not an odd number
    (iii) ☐ is less than fifteen
    (iv) ☐ is not less than seven.

Mathematics Test 4

6—Spatial Discrimination

Which of the following groups of plane shapes could be put together to make solid shapes?

1. 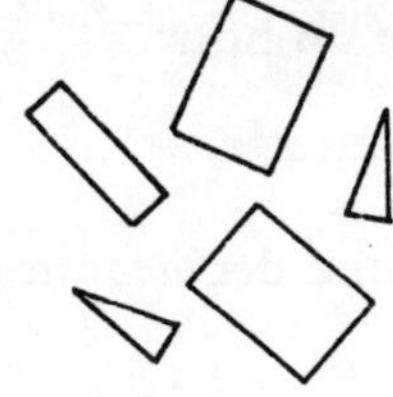

2. 

3. 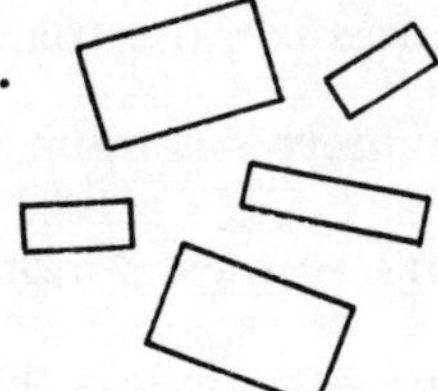

4. 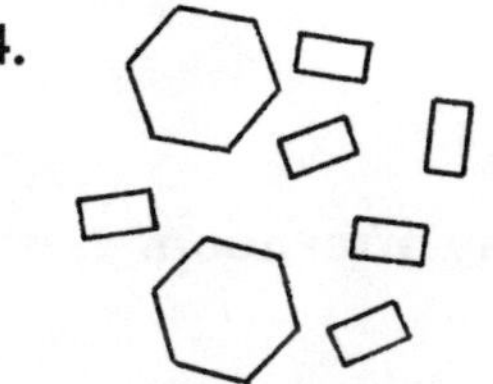

5. 

6. 

Use *one* or *two* of the following words to describe each of the solids you could make in questions 1 to 6:

triangular, square, rectangular, pentagonal, hexagonal, prism, pyramid, octahedron, tetrahedron, cube, cuboid.

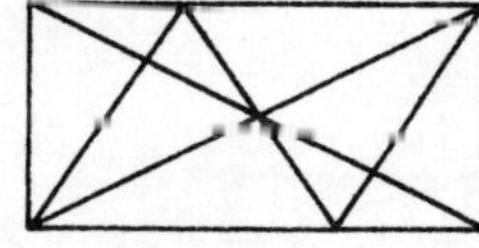

Which of the following shapes are to be found in the figure on the left?

7. 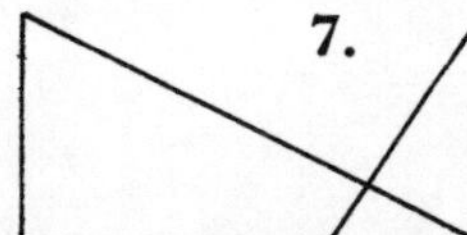
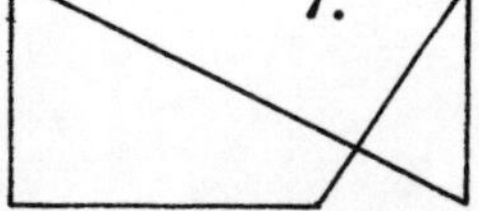

8. 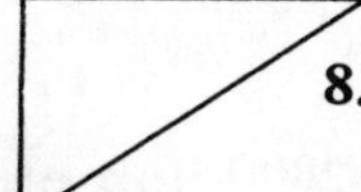

9.

10. 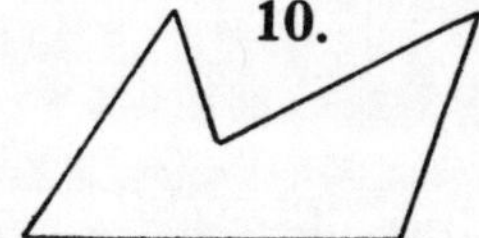

11. 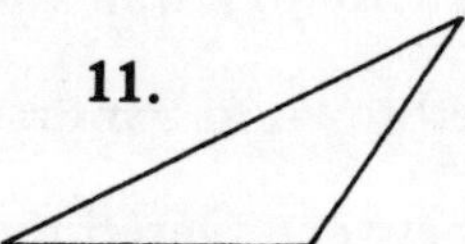

12. 

Mathematics Test 4

7—Measurements and Measuring

Write down one number which would make each of the following statements true. There may be more than one right answer to each question.

1. ☐ minutes is more than 700 seconds but less than a quarter of an hour.
2. ☐ Dekametres is more than 84 metres but less than one Hectametre.
3. ☐ centigrammes is more than 35 milligrammes but less than one decigramme.
4. ☐ centimetres is less than a fifth of a metre.
5. ☐ millilitres is less than one twentieth of a litre.

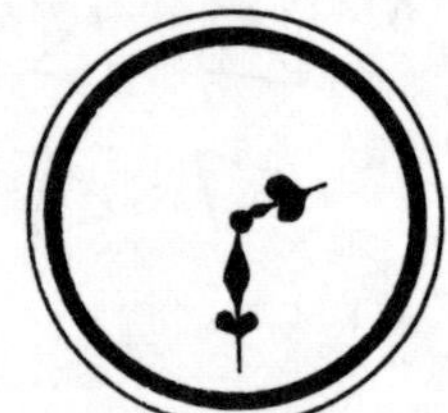

This is the time on a Tuesday afternoon.

6. In how many hours and minutes will it be 4.05 a.m. on Friday?
7. How many hours and minutes have passed since 3.00 p.m. Saturday?

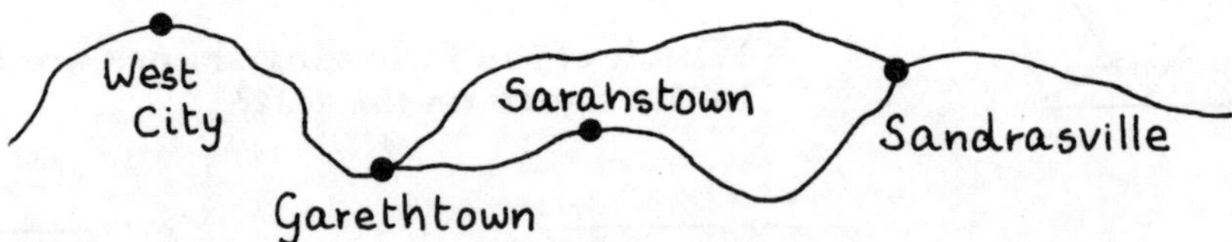

On this map 1 centimetre stands for 200 kilometres.

8. How far is it by road from West City to Sarahstown?
9. How far is it by road from Garethtown to Sandrasville?
10. How far is it 'as the crow flies' between Garethtown and Sandrasville?

Mathematics Test 4

8—Measurements and Measuring

Tony decided to record the speed of his father's car as they went along a stretch of road. Shown below is a graph he drew of his results. Look carefully at the graph and then answer the questions below.

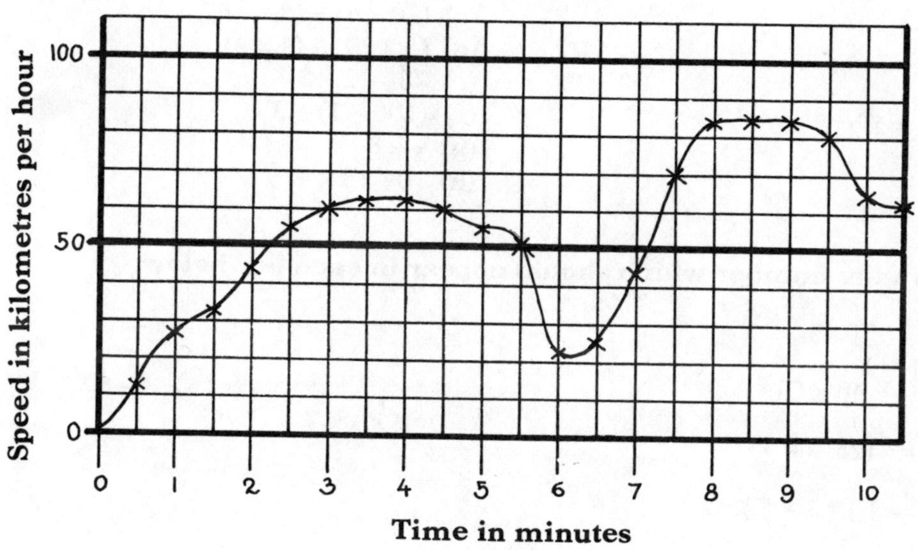

11. How fast was the car going after five minutes?

12. After how many minutes was the car travelling fastest?

13. How many seconds did it take for the car to increase its speed from 10 to 50 kilometres per hour?

14. How many seconds did it take for the car to decrease its speed from 60 to 30 kilometres per hour?

15. For how many seconds was the car travelling faster than 70 kilometres per hour?

16. How many kilometres, approximately, did the car travel during the third minute of Tony's graph?

Mathematics Test 5

1—Basic Computation

For each expression on the left find the one on the right which gives the same answer. Write down its letter.

1. $(7 \times 8) - 23$
2. 7×7
3. 84
4. $11 \times (125 \div 25)$
5. $4 \div 8$

(a) $(100 \div 2) + 8 - 9$
(b) 77
(c) $(20 \times 5) - (9 \times 5)$
(d) $(5 \times 20) - (4 \times 4)$
(e) $78 - (2 \times 3)$
(f) $(100 - 1) \div 3$
(g) $4 - 8$
(h) $10 \div (5 \times 4)$

Write down the number which should appear in each box below.

6. $11 \times 9 = \square \times 3$
7. $12 \times 10 = 40 \times \square$
8. $8 \times \square = 128 \div 4$
9. $10 \times 5 = \square \times 2 \times \square$
10. $\frac{1}{4} + \frac{3}{16} = 7 \times \square$
11. £3.42 + 28p + £0.18 = $\square$
12. £1.35 = $\square \times$ 9p
13. £1.36 $\div$ 8 = $\square$
14. $\frac{3}{7}$ of £1.68 = $\square$
15. $\frac{5}{6}$ of 84p = $\square$

Mathematics Test 5

2—Understanding of Number

Numbers can be given shapes by using dots like this:  =  = **4**

For each number shape on the left find the one on the right which stands for the same number. Write down its letter.

1. 
2. 
3. 
4. 
5. 

(a) 
(b) 
(c) 
(d) 
(e) 
(f) 
(g) 
(h) 

Write down the *smallest* number in each line below.

6. 8322 3822 2382 2283 2238

7. $\frac{7}{3}$ $\frac{19}{8}$ $\frac{27}{11}$ $\frac{8}{2}$ $\frac{12}{5}$

8. $\frac{12}{3}$ 3.3 $14 \div 2$ $3\frac{1}{4}$ $17 \div 6$

Which are the correct answers to these questions, (a), (b), (c) or (d)?

9. $12 - 18$ (a) -6 (b) 6 (c) 0 (d) 30

10. $13 - 20$ (a) 7 (b) -7 (c) -33 (d) 0

11. **Write in figures the number which has:** three tens, eighteen thousands and four hundreds.

12. **Write down the *largest* number which can be made from the figures:** seven, nine, two and nine.

Mathematics Test 5

3—Understanding of Number

If you are in any doubt about what you are being asked to do in the following questions, you should turn to page 30 in this book where problems *like* the ones below are explained fully.

***Four* small cubes can be joined together to make a *row*.**

***Four* rows can be joined together to make a *square*.**

***Four* squares can be joined together to make a *large cube*.**

23 small cubes will make:

Large Cubes	Squares	Rows	Left Over
	1	1	3

How many large cubes, squares and rows will the following cubes make, and how many cubes will be left over? Write your answers in your book.

	Small Cubes	Large Cubes	Squares	Rows	Left Over
13.	36				
14.	45				
15.	56				
16.	75				
17.	97				

18. If *five* small cubes are used to make a row, how many small cubes will be needed to make a large cube?

19. If 27 small cubes make a large cube, how many are in a row?

Mathematics Test 5

4—Sets, Sequences and Reasoning

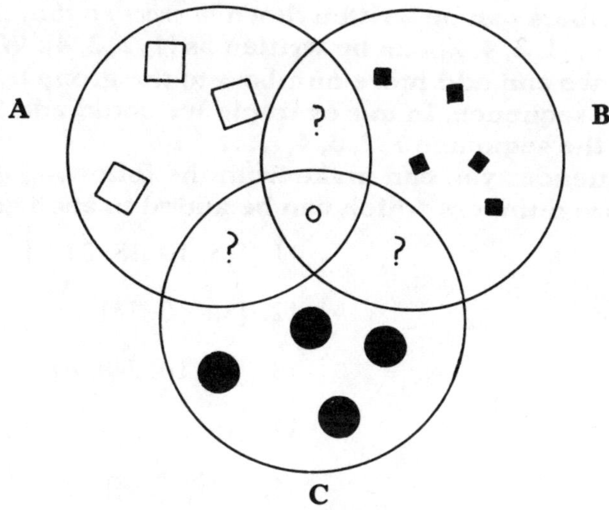

Write down the answers to the following questions:

1. Circle A contains all the shapes that are: (a) large
   (b) white
   (c) square
   (d) black
   (e) round

2. Circle B contains all the shapes that are: (a) round
   (b) black
   (c) square
   (d) small
   (e) large

3. Draw one shape which would be contained by both circle A and circle C, but not by circle B. State whether it is large or small.

4. Draw one shape which would be contained by both circle B and circle C, but not by circle A. State whether it is large or small.

5. When all the shapes in A and C have been taken away the shapes that are left are all: (a) square and black
   (b) square and large
   (c) small and round
   (d) black and round

Mathematics Test 5

5—Sets, Sequences and Reasoning

Some groups of numbers can be written down *in order* so that they form a *sequence*. For example, [1, 3, 4, 2] can be written as [1, 2, 3, 4]. When we have formed a sequence we can add more numbers to the group by working out the next numbers in the sequence. In our example we could add 5 and 6 because these come next in the sequence 1, 2, 3, 4, ...
Write down the sequences you can make from the following groups, and then write down two more numbers which can be added to each group.

6. [56, 52, 64, 60]
7. [45, 38, 52, 31]
8. [34, 48, 62, 20]
9. [8, 216, 72, 24]
10. [10, 12, 7, 5, 15]
11. [15, 10, 18, 23, 7]
12. [3, 6, 3, 18]
13. [40, 10, 240, 5]
14. [$\frac{3}{5}$, 1, $\frac{2}{5}$, $\frac{8}{5}$]
15. [$\frac{5}{9}$, $\frac{20}{81}$, $\frac{5}{6}$, $\frac{10}{27}$]

Here are some statements about a number:
(i) ☐ is a whole number
(ii) ☐ is an odd number
(iii) ☐ is less than ten
(iv) ☐ is greater than seven.

If the number *nine* is put in each box then *all* the statements become *true*.

Write down the number which will make *all* the statements true in each of the following questions:

16. (i) ☐ is not greater than twelve

    (ii) ☐ is a factor of ninety

    (iii) ☐ has a factor of three

    (iv) ☐ is greater than six.

17. (i) ☐ is less than thirty

    (ii) ☐ is a multiple of four

    (iii) ☐ is not less than thirteen

    (iv) ☐ is a factor of sixty.

Mathematics Test 5

6—Spatial Discrimination

Which of the following groups of plane shapes could be put together to make solid shapes?

1. 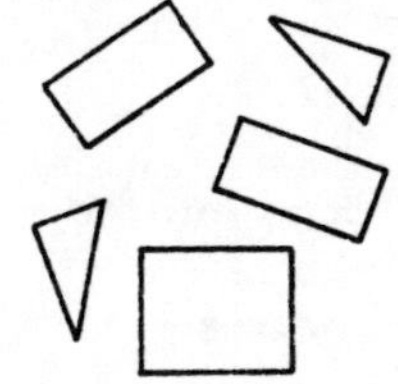

2. 

3. 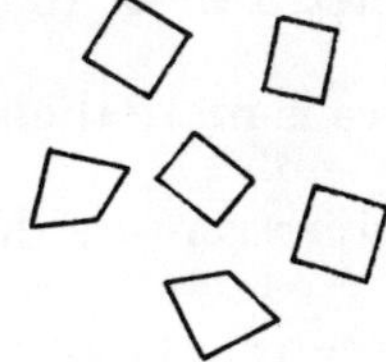

4. 

5. 

6. 

Use *one* or *two* of the following words to describe each of the solids you could make in questions 1 to 6:

triangular, square, rectangular, pentagonal, hexagonal, prism, pyramid, octahedron, tetrahedron, cube, cuboid.

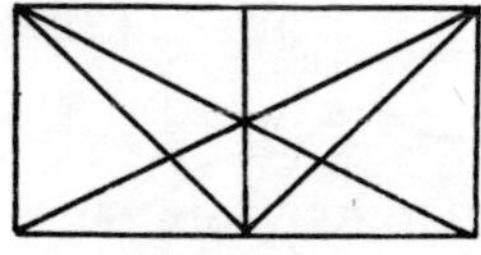

Which of the following shapes are to be found in the figure on the left?

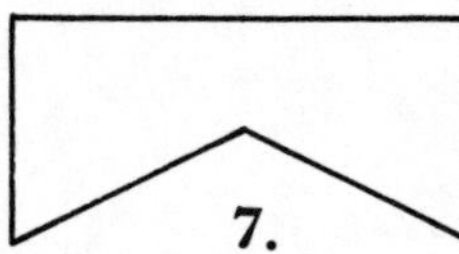

7.

8.

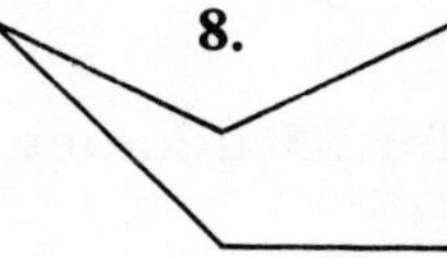

9.

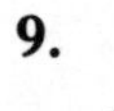

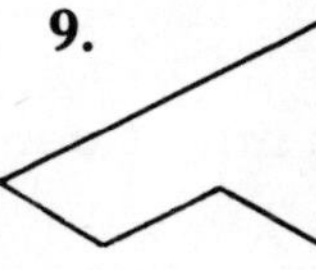

10.

11.

12.

Mathematics Test 5

7—Measurements and Measuring

Write down one number which would make each of the following statements true. There may be more than one right answer to each question.

1. ☐ minutes is more than three hours but less than a sixth of a day.
2. ☐ metres is more than 9 Hectametres but less than one kilometre.
3. ☐ grammes is more than 2 Dekagrammes but less than one Hectagramme.
4. ☐ millimetres is more than 80 centimetres but less than 9 decimetres.
5. ☐ millilitres is less than one fiftieth of a litre.

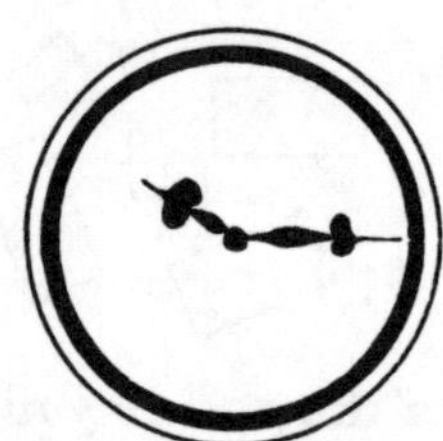

This is the time on a Thursday morning.

6. In how many hours and minutes will it be 9.30 p.m. on Friday?
7. How many hours and minutes have passed since 10.30 p.m. Tuesday?

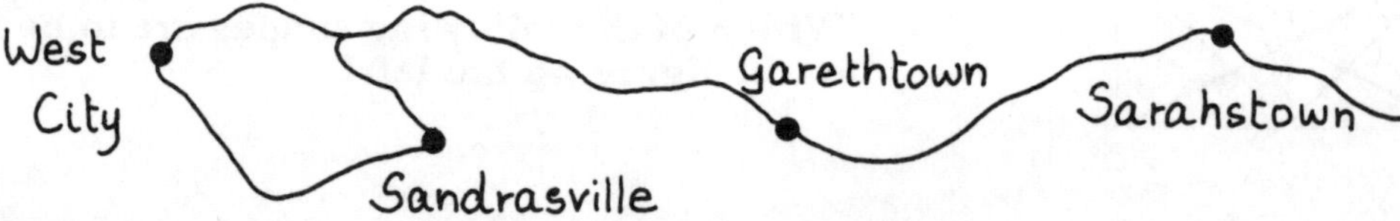

On this map 1 centimetre stands for 500 kilometres.

8. How far is it by road from West City to Garethtown?
9. How far is it by road from Sandrasville to Sarahstown?
10. How much farther is it 'as the crow flies' between Sarahstown and West City than it is between Garethtown and Sarahstown?

Mathematics Test 5

8—Measurements and Measuring

Gareth decided to measure his rate of pedalling as he tried out his new bicycle. Shown below is a graph he drew from his results. Look carefully at the graph and then answer the questions below.

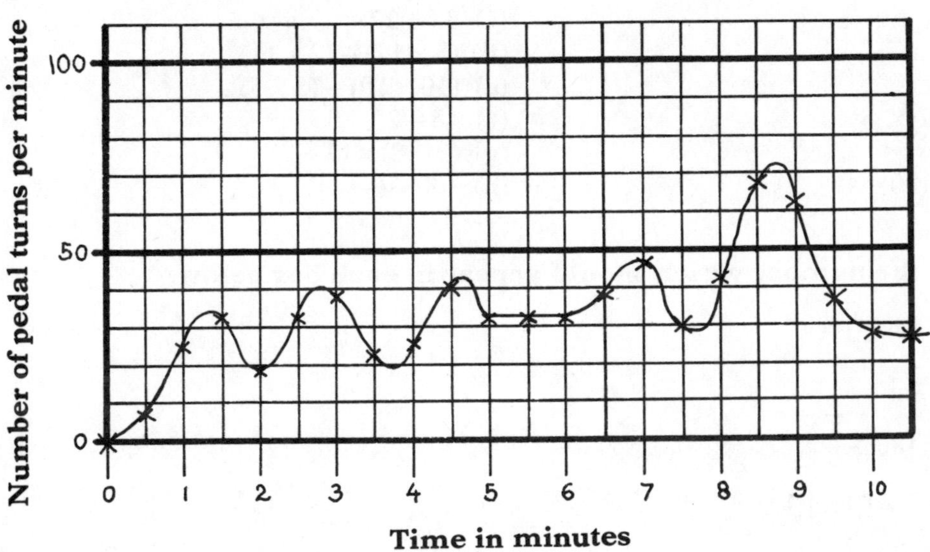

11. What was Gareth's rate of pedalling after three minutes?

12. After how many minutes was Gareth pedalling fastest?

13. How many seconds did it take for Gareth's rate of pedalling to increase from 30 to 60 turns per minute?

14. How many seconds did it take for Gareth's rate of pedalling to decrease from 40 to 20 turns per minute?

15. For how many seconds was Gareth's rate of pedalling faster than 40 turns per minute?

16. How many times did Gareth turn his pedals during the sixth minute of his journey?

Mathematics Test 6

1—Basic Computation

For each expression on the left find the one on the right which gives the same answer. Write down its letter.

1. 12×7
2. 8^2
3. $22 \div 18$
4. $(9 \times 8) + (3 \times 2)$
5. $(8 \times 8) \div (20 - 4)$

(a) $(7 \times 4) + (400 \div 8)$
(b) 12^{7}
(c) $33 \div 27$
(d) $(5 \times 12) + (2 \times 11)$
(e) $(10 \times 10) \div (5 \times 5)$
(f) $18 \div 22$
(g) $(9+8) \times (3+2)$
(h) $100 - (4 \times 4)$

Write down the number which should appear in each box below.

6. $13 \times 6 = 3 \times \square$
7. $5 \times 15 = \square \times 3$
8. $9 \times \square = 180 \div 4$
9. $35 \times 3 = 7 \times \square \times \square$
10. $\frac{3}{8} + \frac{5}{24} = \frac{2}{3} \times \square$
11. £2.13 + 49p + £0.28 = $\square$
12. £1.92 = $\square$ × 12p
13. £1.52 ÷ 8 = $\square$
14. $\frac{4}{9}$ of £1.08 = $\square$
15. $\frac{2}{5}$ of £2.35 = $\square$

Mathematics Test 6

2—Understanding of Number

Numbers can be given shapes by using dots like this: ⁞ = ⁞ = 5

For each number shape on the left find the one on the right which stands for the same number. Write down its letter.

1. ⁞ (a) ⁞ (e) ⁞

2. ⁞ (b) ⁞ (f) ⁞

3. ⁞ (c) ⁞ (g) ⁞

4. ⁞ (d) ⁞ (h) ⁞

5. ⁞

Write down the *smallest* number in each line below.

6. 3285 5283 8532 3258 3528

7. $\frac{5}{4}$ $\frac{7}{6}$ $\frac{9}{8}$ $\frac{3}{2}$ $\frac{17}{16}$

8. $18 \div 7$ 2.87 $19 \div 6$ $2\frac{1}{2}$ $\frac{22}{8}$

Which are the correct answers to these questions, (a), (b), (c) or (d)?

9. $10-20$ (a) 10 (b) -10 (c) 0 (d) 30

10. $5-17$ (a) -12 (b) 0 (c) 12 (d) -22

11. **Write in figures the number which has:** nineteen thousands, eight units and three hundreds.

12. **Write down the *largest* number which can be made from the figures:** three, seven, three, eight and five.

Mathematics Test 6

3—Understanding of Number

If you are in any doubt about what you are being asked to do in the following questions, you should turn to page 30 in this book where problems *like* the ones below are explained fully.

Seven **small cubes can be joined together to make a** *row.*

Seven **rows can be joined together to make a** *square.*

Seven **squares can be joined together to make a** *large cube.*

40 small cubes will make:

Large Cubes	Squares	Rows	Left Over
		5	5

How many large cubes, squares and rows will the following cubes make, and how many cubes will be left over? Write your answers in your book.

	Small Cubes	Large Cubes	Squares	Rows	Left Over
13.	89				
14.	110				
15.	74				
16.	122				
17.	403				

18. If *six* small cubes are used to make a row, how many small cubes will be needed to make a large cube?

19. If 512 small cubes are in a large cube, how many are in a row?

Mathematics Test 6

4—Sets, Sequences and Reasoning

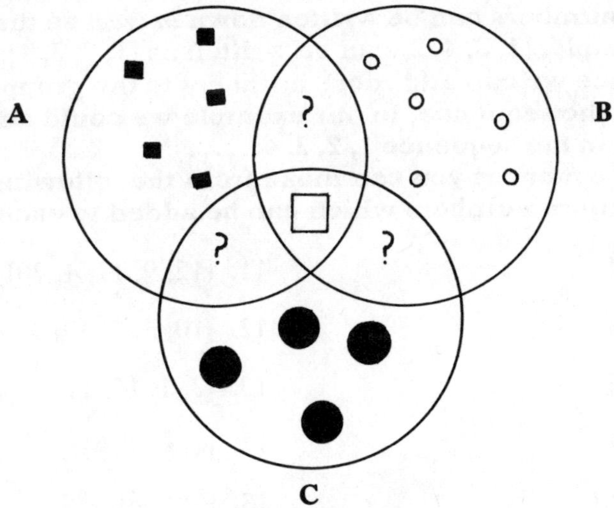

Write down the answers to the following questions:

1. Circle A contains all the shapes that are: (a) small
   (b) black
   (c) square
   (d) large
   (e) white

2. Circle B contains all the shapes that are: (a) round
   (b) white
   (c) small
   (d) square
   (e) large

3. Draw one shape which would be contained by both circle A and circle C, but not by circle B. State whether it is large or small.

4. Draw one shape which would be contained by both circle B and circle C, but not by circle A. State whether it is large or small.

5. When all the shapes in circles B and C have been taken away the shapes that are left are all: (a) black and round
   (b) small and white
   (c) square and large
   (d) small and square

Mathematics Test 6

5—Sets, Sequences and Reasoning

Some groups of numbers can be written down *in order* so that they form a *sequence*. For example, [1, 3, 4, 2] can be written as [1, 2, 3, 4]. When we have formed a sequence we can add more numbers to the group by working out the next numbers in the sequence. In our example we could add 5 and 6 because these come next in the sequence 1, 2, 3, 4,.....
Write down the sequences you can make from the following groups, and then write down two more numbers which can be added to each group.

6. [81, 76, 91, 86]
7. [29, 21, 37, 13]
8. [43, 59, 27, 75]
9. [108, 12, 4, 36]
10. [10, 17, 13, 3, 6]
11. [12, 9, 17, 4, 20]
12. [10, 5, 5, 30]
13. [3, 1, 15, 1]
14. [1, $\frac{5}{7}$, $\frac{10}{7}$, $\frac{4}{7}$]
15. [$\frac{3}{5}$, 1, $\frac{27}{125}$, $\frac{9}{25}$]

Here are some statements about a number:
(i) ☐ is a whole number
(ii) ☐ is an odd number
(iii) ☐ is less than ten
(iv) ☐ is greater than seven.

If the number *nine* is put in each box then *all* the statements become *true*.

Write down the number which will make *all* the statements true in each of the following questions:

16. (i) ☐ is not greater than twenty
    (ii) ☐ has a factor of seven
    (iii) ☐ is a factor of fifty-six
    (iv) ☐ is not less than ten.
17. (i) ☐ is a multiple of two
    (ii) ☐ is a factor of fifty-four
    (iii) ☐ is not less than ten
    (iv) ☐ is not greater than twenty.

Mathematics Test 6

6—Spatial Discrimination

Which of the following groups of plane shapes could be put together to make solid shapes?

1.

2.

3.

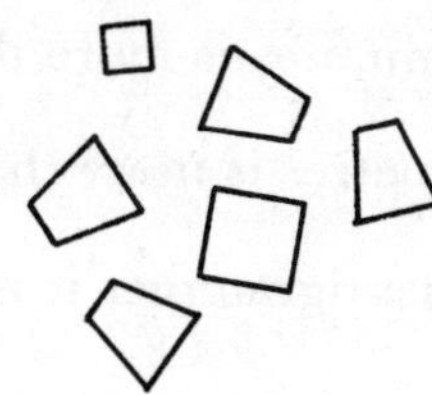

4.

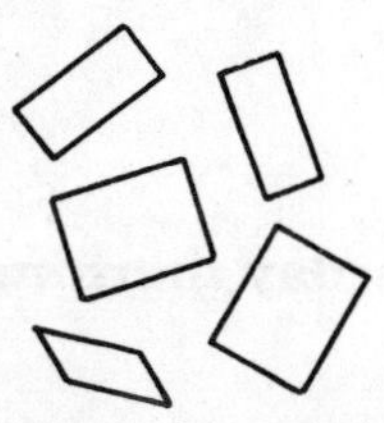

5.

6.

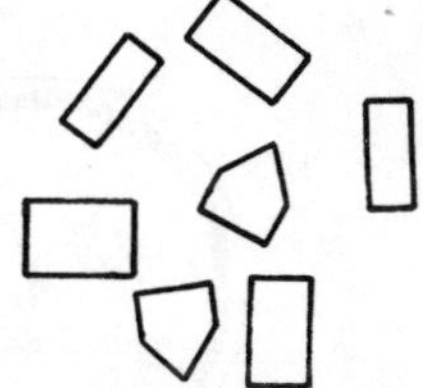

Use *one* or *two* of the following words to describe each of the solids you could make in questions 1 to 6:

triangular, square, rectangular, pentagonal, hexagonal, truncated, prism, pyramid, octahedron, cube, cuboid.

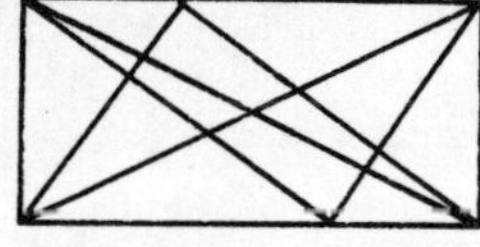

Which of the following shapes are to be found in the figure on the left?

7.

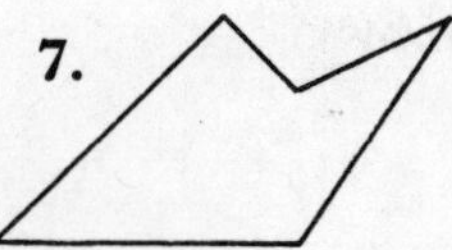

8.

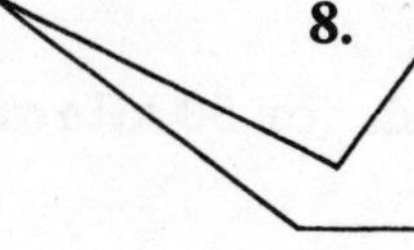

9.

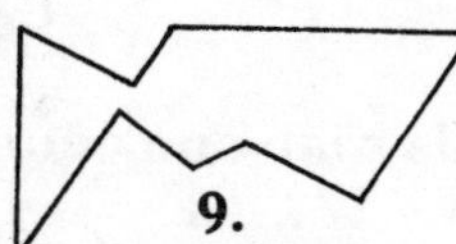

10.

11.

12.

Mathematics Test 6

7—Measurements and Measuring

Write down one number which would make each of the following statements true. There may be more than one right answer to each question.

1. ☐ minutes is more than 700 seconds but less than half an hour.
2. ☐ metres is more than 9 Dekametres but less than one Hectametre.
3. ☐ centigrammes is more than 400 milligrammes but less than half a gramme.
4. ☐ decimetres is more than 60 centimetres but less than one metre.
5. ☐ cubic centimetres is more than one cubic metre.

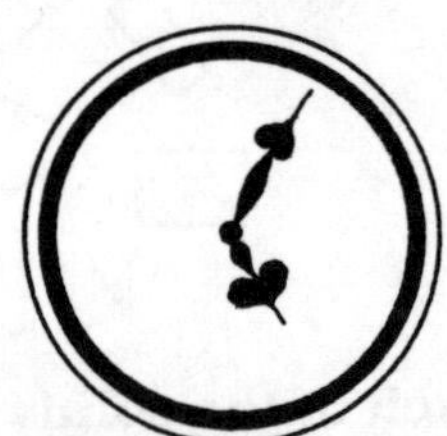

This is the time on a Saturday morning.

6. In how many hours and minutes will it be 4.30 p.m. on Monday?
7. How many hours and minutes have passed since 5.30 p.m. on Thursday?

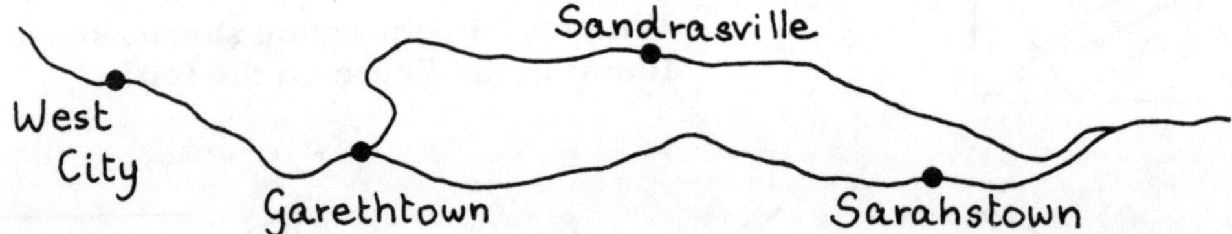

On this map one centimetre stands for 50 kilometres.

8. How far is it by road from West City to Sandrasville?
9. How far is it by road from Garethtown to Sarahstown?
10. How much farther is it 'as the crow flies' between Sarahstown and West City than it is between Garethtown and Sandrasville?

Mathematics Test 6

8—Measurements and Measuring

Sandra decided to measure Gareth's rate of breathing whilst he did his football training. Shown below is a graph she drew of her results. Look carefully at the graph and then answer the questions below.

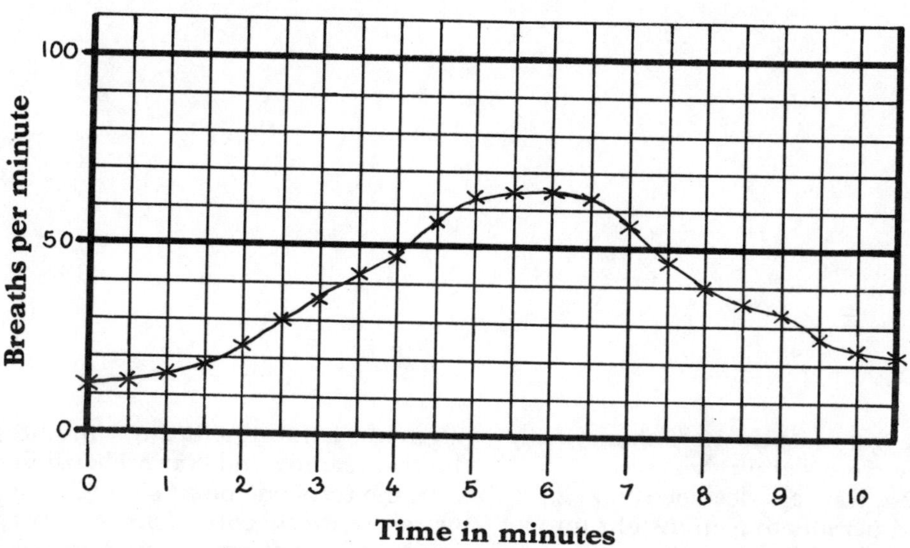

11. What was Gareth's rate of breathing after three minutes?

12. After how many minutes was Gareth's breathing fastest?

13. How many seconds did it take for Gareth's rate of breathing to increase from 20 to 40 breaths per minute?

14. How many seconds did it take for Gareth's rate of breathing to decrease from 60 to 30 breaths per minute?

15. For how many seconds was Gareth's rate of breathing faster than 35 breaths per minute?

16. How many breaths, approximately, did Gareth take during the sixth minute of his training session?

A Guide to the English Usage Tests

An effective test must do more than merely evaluate one child's ability vis-à-vis others. These tests are designed to help teachers and parents to gain insight into a child's particular difficulties. They can thus be used as a springboard for further learning.

We therefore strongly urge that time be made available for post-test discussions. These should concentrate on the strategics of thinking employed by the child. Often he or she 'knows' the correct answer, but is unable to utilize this knowledge. Merely to concentrate on pointing out mistakes is negative and could be psychologically crippling.

These tests aim at above-average standards; easy English tests, which everyone completes successfully, test no more than handwriting.

With regard to timing, we suggest that fifty minutes be allowed for the first four pages of each test, and thirty minutes for the two reading and comprehension pages.

In the tests one point should be given for each entirely correct answer. If two words instead of one are underlined, or more than one answer is given, no points are gained. The total marks for each complete test amount to eighty points.

The spelling and punctuation exercises are based on expectations of standards achieved by able ten- to eleven-year-olds.

The sentence-construction exercises are inevitably open to differing answers. The inventive child will, no doubt, be able to produce alternatives. As long as these are sensible, and the child can justify them, they should be accepted.

The answers to the first page of the reading and comprehension tests are more open-ended, and marking (3 points maximum each answer) is left to the discretion of the teacher or parent.

English Usage Test 1

1—Spelling and Punctuation

***One* of the words in each of the following sentences has been mis-spelt. Write out each mis-spelt word correctly.**

1. 'Do you rearly like chocolate?' said Gareth.
2. Do you enjoy histry?
3. We use washing-up licwid.
4. Sandra went to musick lessons every Saturday morning.
5. Joe's hair was of mediam length.
6. Sarah's favourite subjects were geography and mathmatics.
7. Would you like a Spanish Omerlet?
8. We have twelve mirrers in our flat.
9. Gareth was exsorsted after the four mile walk.
10. Sarah spilt a cup of coffee on the new carpit.

***Punctuate* the following sentences:**

1. the river thames flows through london
2. for supper we cooked sausages chips and beans
3. is there enough food for breakfast asked sarah
4. dont do that
5. all joes trousers were levis

English Usage Test 1

2—Vocabulary: synonyms

Write down the word in the second sentence which means the *same* as the word underlined in the first sentence.

1. It feels awful to take home a bad school report.
   I ought to report this terrible accident.

2. Salt water makes you feel sick.
   Sallow skin can make you look ill.

3. I inquired where the cloakroom was.
   'What is a quire?' asked Sandra.

4. Although it looked peculiar, it was the latest dance.
   Looking at the fashionable shops made Sarah late.

5. Joe groaned when he looked at the homework.
   'Groats for breakfast, again!' moaned Joe.

By writing *one* word in each of the spaces, make the second sentence mean the *same* as the first.

1. Games are compulsory in our school.
   In school (   ) (   ) do games.

2. The children rejoiced at winning the game.
   They were (   ) to have (   ) (   ) (   ).

3. Gareth did not forget the birthday card.
   Gareth (   ) the birthday card.

4. It was with trembling hands that she carried the vase.
   (   ) the vase (   ) (   ) (   ) shake.

5. I enjoy your cooking.
   I (   ) (   ) cooking (   ) much.

English Usage Test 1

3—Vocabulary: antonyms

Write down the word in the second sentence that is *opposite* to the word underlined in the first sentence.

1. She emerged from the sea covered in seaweed.
   It was the submerged wreck of a merchant ship.
2. We concealed her birthday presents under the bed.
   The garden was conveniently exposed to the sun all day.
3. The children bought ten comics between them.
   Sarah brought her old toys, to be sold in school.
4. I would love to be two inches taller.
   Girls like to be kissed, but hate to be hugged.
5. Camping out was a tremendous success.
   Sandra's 'fish' lunch was a failure and a mess.

If you were writing a story called 'A Night at Green Mansions', in what order would you use the following headings?

A Night at Green Mansions

See ghostly hand reappear.

Faint with fright.

The hand disappears and a skull appears.

Waking and seeing a hand on the pillow.

Going up to bed.

English Usage Test 1

4—Sentence Construction

Form five different sentences by crossing out one or more of the words used in the sentence below. The first has been done for you.

The astronauts received and signalled messages.
The astronauts ~~received and~~ signalled messages.

Form the following groups of words into sentences by re-arranging them and adding *one* of your own.

1. cricket first Gareth the eleven played
2. match football first scored Joe in the goal
3. had had Sarah party it birthday best the ever
4. shorts wore knee- referee the
5. snap Sandra the game play to hated

Do exactly the same for the next groups of words, only this time some of the words have been placed in order for you. Remember to add one word of your own.

6. day all Sunday last rained
   ( ) ( ) ( ) rained ( ) day.
7. out fall caused filling Gareth's toffee the
   The ( ) ( ) Gareth's ( ) ( ) fall ( ).
8. Easter for country children the went to
   ( ) children ( ) to ( ) ( ) for ( ).
9. crying she was I did know not whether laughing
   I did ( ) ( ) whether ( ) was ( ) ( ) crying.
10. hair fair blondes
    Blondes ( ) ( ) hair.

English Usage Test 1

5—Reading and Comprehension

Read the following passage carefully, and then answer the questions below.

'Now breakfast,' said Merlyn.

The Wart saw that the most perfect breakfast table was laid out neatly for two, on a table before the window. There were peaches. There were also melons, strawberries and cream, rusks, brown trout piping hot, grilled perch which were much nicer, chicken devilled enough to burn one's mouth out, kidneys and mushrooms on toast, fricassee, curry, and a choice of boiling coffee and best chocolate made with cream in large cups.

'Have some mustard,' said Merlyn, when they got to the kidneys.

The mustard-pot got up and walked over to his plate on thin silver legs that waddled like an owl's. Then it uncurled its handles, and one handle lifted its lid with exaggerated courtesy, while the other helped him a generous spoonful.

'Oh, I love the mustard-pot!' cried the Wart.

'Where did you get it?'

At this, the pot beamed all over its face, and began to strut a bit; but Merlyn rapped it on the head with a teaspoon, so that it sat down and shut up at once.

'It's not a bad pot,' he said grudgingly, 'only it is inclined to give itself airs.'

1. **What fish did the Wart prefer?**
2. **What was the mustard for?**
3. **What was the mustard-pot compared with?**
4. **Why did Merlyn smack the pot?**
5. **What do the following words mean: (a) perfect**
   (b) exaggerated
   (c) generous?

English Usage Test 1

6—Reading and Comprehension

In the following passage, several words have been left out. The first letter of each missing word has been put in to help you. Write down the missing words, in the order in which they occur in the passage.

Camouflage

Colour is one of the main methods of c______. It serves the animal in both attack and defence. If an animal remains perfectly still in its natural habitat, it will remain unseen by its e______. It also allows an animal to remain u______ until it gets within striking distance of its prey. Frogs and toads are c______ green to resemble the marshy areas which they inhabit, while arctic creatures are snowy white, r______ their h______.

Creatures which i______ the sea have many different methods of camouflage. They may a______ pieces of sponge and weed to their bodies, or be able to eject a cloud of ink at their enemy, so that they may swim away, in the now-darkened waters.

One of the most extraordinary methods of camouflage is the ability some creatures have of changing their c______. A chameleon is the most well-known example of these c______. It can begin crossing a dusty, reddish road, green colour to match the green verge. As it m______ across the road, it becomes less bright and turns to a r______ brown. On reaching the other grass v______ it begins o______ again to turn green.

C______ gives the weakest animals a chance to survive, and also adds to the efficiency of the hunter after its prey.

English Usage Test 2

1—Spelling and Punctuation

***One* of the words in each of the following sentences has been mis-spelt. Write out each mis-spelt word correctly.**

1. Romeo hid beneath the balcony, by a large piller.
2. Our green-groser sells very cheap fruit.
3. Joe's father keeps home-made wine in their celler.
4. Centrally heated houses rarely have big chimnees.
5. 'Don't menshon that creep to me!' cried Sarah.
6. My uncle keeps parrats as pets.
7. Gareth's hankerchiefs are always crumpled.
8. What is the capitle city of Paraguay?
9. Falling from the wall-bars gave Sandra a bad headake.
10. Joe's favourite vegeterble is cabbage.

***Punctuate* the following sentences:**

1. sarahs teeth were the whitest in the class
2. how much pocket money do you get
3. please be quiet said miss roberts
4. we put the paints brushes books and crayons in the trunk
5. name the country nearest to new zealand

English Usage Test 2

2—Vocabulary: synonyms

Write down the word in the second sentence which means the *same* as the word underlined in the first sentence.

1. Sarah persuaded Joe to buy her an ice-cream.
   I am convinced that the earth is not perfectly round.
2. Jason went in search of the legendary Golden Fleece.
   The tyres screeched in the car's pursuit of the robbers.
3. It is not very difficult to connect the wires to a plug.
   The spy contacted his chief link.
4. All scholars are not bookworms.
   In the new school the pupils wore scarlet scarves.
5. Joe couldn't remember how to subtract seven from two.
   Deduct the club subscription from your pocket money.

By writing *one* word in each of the spaces, make the second sentence mean the *same* as the first.

1. The flat was burgled in her absence.
   While (   ) was (   ) the flat (   ) (   ).
2. He turned away as if he had not heard.
   He (   ) not to (   ) (   ).
3. I dislike jellybabies intensely.
   (   ) (   ) jellybabies.
4. The clock struck twelve o'clock at night.
   (   ) clock (   ) (   ).
5. Sandra was going the same way as myself.
   Sandra was going (   ) (   ).

English Usage Test 2

3—Vocabulary: antonyms

Write down the word in the second sentence that is *opposite* to the word underlined in the first sentence.

1. Gareth proved very fit in his recent games.
   Joe collected old American ten-cent comics.

2. A fragment from the broken vase stuck in Sarah's thumb.
   Sarah planted the whole of her plot with fragrant flowers.

3. Attached to his jeans was a large, leather label.
   The top of the egg was detached by an attack with a knife.

4. We covered the rear window of our car with stickers.
   Our hotel was near the sea front.

5. Buying a dozen roses in winter was very generous.
   I felt mean in giving her such a small meal.

If you were writing a story called 'A Day's Fishing', in what order would you use the following headings?

A Day's Fishing

Float pulled under water.

Baiting the line.

Telling friend about the one which got away.

Enormous fish escapes off hook.

Digging for very fat, juicy worms.

English Usage Test 2

4—Sentence Construction

Form five different sentences by crossing out one or more of the words used in the sentence below. The first has been done for you.

When will she have time to do it?
When will she ~~have time to~~ do it?

Form the following groups of words into sentences by re-arranging them and adding *one* of your own.

1. foreign Polish is a
2. gold valuable why?
3. the alphabet of what the is twelfth?
4. hour an sixty are there in
5. trout went we

Do exactly the same for the next groups of words, only this time some of the words have been placed in order for you. Remember to add one word of your own.

6. born Scorpio of Joe was sign the
   Joe ( ) ( ) ( ) the sign ( ) ( ).
7. horoscopes believe you do?
   Do ( ) ( ) ( ) horoscopes?
8. god underworld Greek of Hades the was
   ( ) was the ( ) ( ) of ( ) ( ).
9. filled coffee chocolate was the
   ( ) chocolate ( ) ( ) ( ) coffee.
10. marry him promised she
    She ( ) ( ) marry ( ).

English Usage Test 2

5—Reading and Comprehension

Read the following passage carefully, and then answer the questions below.

'Look'—cried Peter suddenly—'the tree over there!'

The tree he pointed at was one of those that have rough grey leaves and white flowers. The berries, when they come, are bright scarlet, but if you pick them, they disappoint you by turning black before you get them home. And, as Peter pointed, the tree was moving—not just the way trees ought to move when the wind blows through them—but all in one piece, walking down the side of the cutting.

'It's moving!' cried Bobbie. 'Oh, look! and so are the others. It's like the woods in Macbeth.'

'It's magic!' said Phyllis, breathlessly. 'I always knew the railway was enchanted.'

It really did seem a little like magic. For all the trees for about twenty yards of the opposite bank seemed to be slowly walking down towards the railway line, the tree with the grey leaves bringing up the rear, like some old shepherd driving a flock of sheep.

'What is it? Oh, what is it?' said Phyllis. 'It's much too magic for me, I don't like it. Let's go home.'

But Bobbie and Peter clung fast to the rail and watched breathlessly. And Phyllis made no movement towards going home by herself.

The trees moved on and on. Some stone and loose earth fell down, and rattled on the railway metals far below.

1. **Who saw the trees move first?**
2. **How did the trees move?**
3. **Why did Phyllis want to go home?**
4. **What was the grey-leafed tree compared with?**
5. **What do the following words mean: (a) enchanted**
   (b) disappoint
   (c) opposite?

English Usage Test 2

6—Reading and Comprehension

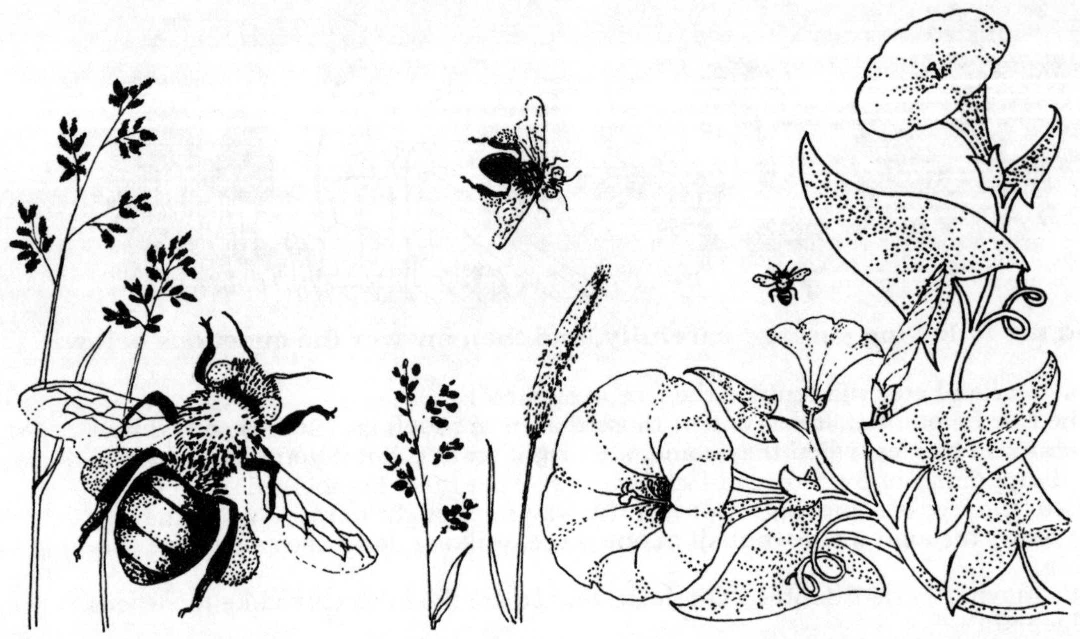

In the following passage, several words have been left out. The first letter of each missing word has been put in to help you. Write down the missing words, in the order in which they occur in the passage.

The Honey-bee

In a bee-hive community, there are three types of b______: drones, workers and one queen-bee. The w______ are the only bees of the c______ which visit flowers and collect the nectar with which to make h______.

To reach the n______ the honey-bee has an extremely long tongue covered with tiny hairs. It gathers the minute drops of nectar up with a sweep of its h______ tongue, and stores it up in a type of stomach called a honey-sac. While the bee is still c______ nectar from other flowers, the nectar on the honey-s______ is being changed by chemicals into h______. When the bee becomes h______ it can open a tiny valve in the honey-sac, and allow some honey to trickle into its own stomach as f______.

The nectar from more than three hundred f______ makes only one drop of honey. When the worker's sac is full, the bee d______ it b______ to the h______.

English Usage Test 3

1—Spelling and Punctuation

***One* of the words in each of the following sentences has been mis-spelt. Write out each mis-spelt word correctly.**

1. Can you sirgest a game for two players?
2. The paper seemed to have anchent writing on it.
3. 'That's a very perculier hair-cut, Joe,' giggled Sarah.
4. The chocolates disappeared in a very mistarious way.
5. There is genrelly a good match on Saturdays.
6. We don't sing himz at assembly any more.
7. Joe finully learnt how to head a ball.
8. There were gards and dogs the length of the wall.
9. The children went threw their homework together.
10. The mouse nawed a huge hole.

***Punctuate* the following sentences:**

1. shall we buy him a toy sweets or a book
2. may i borrow your gun asked Joe
3. sarah supported manchester united
4. shes got blond hair
5. gareths hair was brown

English Usage Test 3

2—Vocabulary: synonyms

Write down the word in the second sentence which means the *same* as the word underlined in the first sentence.

1. At the end of the term we <u>exchanged</u> addresses.
   The class swot swopped books and changed topics.

2. We had a <u>pact</u> not to tell tales on each other.
   The army promised to sign a treaty with the Indians.

3. Because he was a new boy, he felt <u>uncertain</u> of himself.
   My mother was doubtful about the new kitchen curtains.

4. The most <u>celebrated</u> of all Greek oracles was at Delphi.
   The celestial music of the cellist was world famous.

5. Our cat has <u>sleek</u> black fur.
   You can make your hair glossy by rubbing it with silk.

By writing *one* word in each of the spaces, make the second sentence mean the *same* as the first.

1. While I was away they changed the team.
   During (   ) (   ) they changed the team.

2. Avoid railway lines.
   (   ) (   ) from railway lines.

3. He knew of his team's relegation.
   He knew that (   ) (   ) (   ) been (   ).

4. He put a worm on his hook.
   He (   ) his hook.

5. She put back the books.
   She (   ) the books.

English Usage Test 3

3—Vocabulary: antonyms

Write down the word in the second sentence that is *opposite* to the word underlined in the first sentence.

1. Losing a penny was a trivial matter.
   We shouldn't have laughed at the serious film.
2. We rejected his application to join our club.
   The sovereign accepted the sceptre.
3. We whispered 'Ice-cream' in Sarah's ear.
   'Whisky and cake!' screamed the Earl to the waiter.
4. Sarah retreated into our clubhouse during the storm.
   Using their advantage the troops advanced.
5. The children choose their own clothes.
   The class were given choice books.

If you were writing a story called 'Holiday in a Tent', in what order would you use the following headings?

Holiday in a Tent

Pitching the tent.

Finding a good site.

Back to sleep in van.

Getting rained out.

Slithering into wet sleeping bags.

English Usage Test 3

4—Sentence Construction

Form five different sentences by crossing out one or more of the words used in the sentence below. The first has been done for you.

Don't shout out her name.
~~Don't~~ shout ~~out~~ her name.

Form the following groups of words into sentences by re-arranging them and adding *one* of your own.

1. door a bolt will a key or the?
2. you good milk very is
3. very hard- jeans denim are
4. from milk made cheese
5. Kenneth by wind willows the in was Grahame

Do exactly the same for the next groups of words, only this time some of the words have been placed in order for you. Remember to add one word of your own.

6. Circus is Eros statue Piccadilly of the
   The statue (   ) (   ) is (   ) (   ) (   ).
7. sometimes colour in white violets wild
   Wild (   ) (   ) sometimes (   ) in (   ).
8. regiment rides cavalry a
   A (   ) (   ) rides (   ).
9. covers moon the the an during eclipse
   (   ) an eclipse (   ) moon (   ) (   ) (   ).
10. Flora called flowers goddess Italian the was
    (   ) Italian goddess (   ) flowers (   ) (   ) (   ).

English Usage Test 3

5—Reading and Comprehension

Read the following passage carefully, and then answer the questions below.

Every afternoon, as they were coming from school, the children used to go and play in the Giant's garden.

It was a large lovely garden, with soft green grass. Here and there over the grass stood beautiful flowers, like stars, and there were twelve peach trees that, in the spring-time, broke out into delicate blossoms of pink and pearl, and in the autumn bore rich fruit. The birds sat on the trees, and sang so sweetly that the children used to stop their games in order to listen to them. 'How happy we are here!' they cried to each other.

One day the Giant came back. He had been to visit his friend, the Cornish ogre, and had stayed with him for seven years. After the seven years were over, he had said all that he had to say, for his conversation was limited, and he determined to return to his own castle. When he arrived, he saw the children playing in the garden.

'What are you doing here?' he cried in a very gruff voice, and all the children ran away.

'My own garden is my own garden' said the Giant; 'anyone can understand that, and I will allow nobody to play in it but myself.' So he built a high wall around it, and put up a notice board: TRESPASSERS
WILL BE
PROSECUTED

He was a very selfish Giant.

1. **Describe the garden in detail.**
2. **Why was the Giant selfish?**
3. **Why do you think the children were happy in the garden?**
4. **Why do you think the Giant was limited in his conversation?**
5. **What do the following words mean: (a) selfish**
   (b) trespassers
   (c) prosecuted?

English Usage Test 3

6—Reading and Comprehension

In the following passage, several words have been left out. The first letter of each missing word has been put in to help you. Write down the missing words, in the order in which they occur in the passage.

The Golden Gate

On the West Coast of North America lies the Bay of San Francisco. A mile of water separates the two peninsulas, which form the G______ Gate. The p______ are joined by the longest single-span suspension bridge in the world. Before the b______ was built, traffic coming or going to San Francisco, situated on the southern p______, had to go many miles around the B______ or cross the w______ by ferry.

First, two tall, steel piers were built. The northern p______ was built at the water's edge, but the southern pier had to be constructed 1100 feet from s______. Next, two thick, strong cables were anchored securely in the ground on one shore, passed over the two piers, and a______ in the ground at the opposite side. Each c______ is made up of 61 s______ and each strand has 452 separate wires. The suspender cables joining the two main cables support a six-l______ roadway and wide footpaths.

The large s______ span is 4200 feet in length and 200 feet above water. Two approach s______ of 1125 feet make the o______ length 6450 feet. The bridge was opened in 1936.

English Usage Test 4

1—Spelling and Punctuation

***One* of the words in each of the following sentences has been mis-spelt. Write out each mis-spelt word correctly.**

1. The train carriges were terribly dirty.
2. Sarah had a breef stay in hospital.
3. Gareth was never jellus of his sister.
4. Joe tried to make his mussels bigger.
5. The whole school went to the sirkus.
6. The children went to the sea-side quite resently.
7. Both girls had similer shaped noses.
8. A peese of the jig-saw puzzle was missing.
9. Sandra bought an oringe sweater.
10. Sinbad, the eagel, had a bald head.

***Punctuate* the following sentences:**

1. who is your favourite singer
2. sarahs knee was badly bruised
3. hes got lots of freckles said sandra
4. she whispered across to joe do you want sarah to come
5. you will marry a tinker tailor soldier or sailor

English Usage Test 4

2—Vocabulary: synonyms

Write down the word in the second sentence which means the *same* as the word underlined in the first sentence.

1. The <u>base</u> of the clock was covered with felt.
   Paint the top, back, bottom and sides of the hut.

2. The children were <u>grouped</u> together by ages.
   We had to grope for a combined athletic team.

3. It was by great <u>cunning</u> that Chivers beat Banks.
   School children were once caned for slyness.

4. I <u>heard</u> a new pop group last night.
   The farmer listened to the herd lowing in the meadow.

5. The colonel <u>rejected</u> any offer of help.
   They refused to eject him from the plane.

By writing *one* word in each of the spaces, make the second sentence mean the *same* as the first.

1. You look like your father.
   You (   ) your father.

2. 'Extinguish the fire' ordered our leader.
   Our leader ordered us to (   ) (   ) the fire.

3. It was out of the question for her to lie.
   She found it (   ) to lie.

4. The actress took advantage of her beauty.
   The actress (   ) her beauty.

5. She looked up to her brother.
   She (   ) her brother.

English Usage Test 4

3—Vocabulary: antonyms

Write down the word in the second sentence that is *opposite* to the word underlined in the first sentence.

1. He was very excitable before his birthday.
   The sea-crossing was exceedingly calm for winter.
2. Have you been on a foreign holiday?
   The native flowers of this country are roses.
3. The entire nation felt happy at the news.
   Part of the class was miserable at having holiday work.
4. It was unusual for Sandra to have nightmares.
   A common hobby was collecting used stamps.
5. What is a just punishment for cheating?
   It is unfair to always send the girls out first.

If you were writing a story called 'School Play', in what order would you use the following headings?

School Play

Curtain up! First night.

Final dress rehearsal.

Rehearsals after school.

Applause and shouts for more.

Auditioning for parts in school play.

English Usage Test 4

4—Sentence Construction

Form five different sentences by crossing out one or more of the words used in the sentence below. The first has been done for you.

Hector and the Trojans planned the attack.
Hector and the Trojans planned ~~the attack~~.

Form the following groups of words into sentences by re-arranging them and adding *one* of your own.

1. watching like I Saturdays television
2. beauty Troy her famous Helen of was
3. leaves Beech trees Copper red very
4. habit your a nails biting
5. satchel carry your a do books you?

Do exactly the same for the next groups of words, only this time some of the words have been placed in order for you. Remember to add one word of your own.

6. year extra has day fourth every
   Every (   ) (   ) has (   ) (   ) (   ).
7. parrots good sunflowers for seeds
   (   ) seeds (   ) good (   ) (   ).
8. Gareth log river by cabin the Joe a built
   (   ) (   ) Joe built (   ) (   ) (   ) (   ) the (   ).
9. few mathematics mistakes made Sarah
   (   ) made (   ) (   ) (   ) mathematics.
10. have tigers white ears on spots
    (   ) have (   ) (   ) on (   ) (   ).

English Usage Test 4

5—Reading and Comprehension

Read the following passage carefully, and then answer the questions below.

Everyone who reads this, knows what it is like to go in a train, so I shall not tell about it—though it was rather fun, especially the part where the guard came for the tickets at Waterloo, and H.O. was under the seat, and pretended to be a dog without a ticket. We went to Charing Cross, and we just went round to Whitehall to see the soldiers, and then by St. James for the same reason—and when we'd looked in the shops a bit, we got to Brook Street, Bond Street. It was a brass plate on a door next to a shop—a very grand place where they sold bonnets and hats—all very bright and smart, and no tickets on them to tell you the price. We rang a bell and a boy opened the door, and we asked for Mr. Rosenbaum. The boy was not polite; he did not ask us in. So then Dicky gave him his visiting card; it was one of Father's really, but the name is the same, Mr. Richard Bastable, and we others wrote our name underneath. I happened to have a piece of pink chalk in my pocket, and we wrote them with that.

Then the boy shut the door in our faces, and we waited on the step. But presently he came down and asked our business, so Dicky said—'Money advanced, young shaver! and don't be all day about it!'

1. **Why did the children go to St. James?**
2. **Why did they think the boy was rude?**
3. **What was Dicky's full name?**
4. **What did the children go to Mr. Rosenbaum for?**
5. **What do the following words mean: (a) especially**
   (b) smart
   (c) polite?

English Usage Test 4

6—Reading and Comprehension

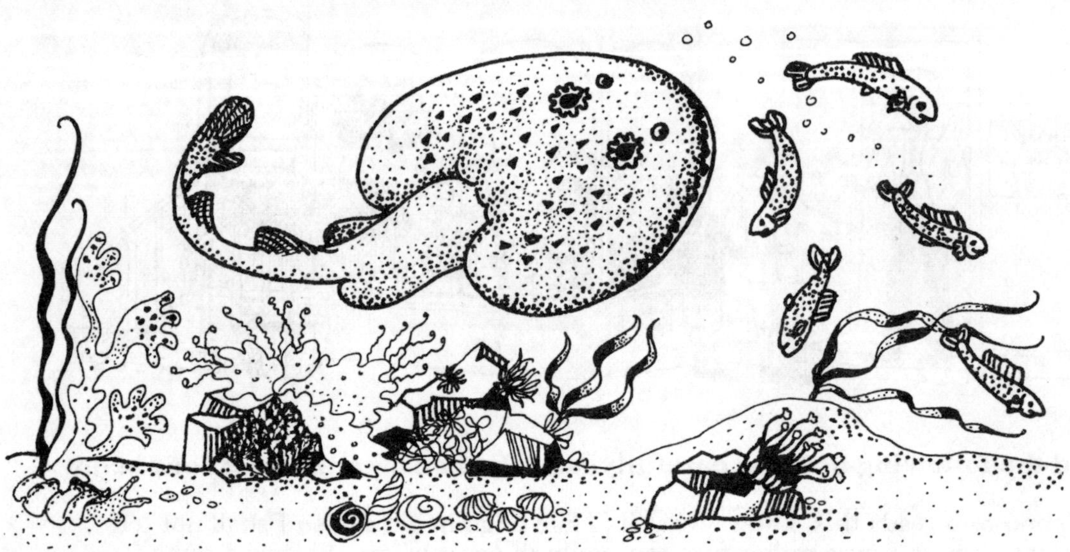

In the following passage, several words have been left out. The first letter of each missing word has been put in to help you. Write down the missing words, in the order in which they occur in the passage.

Electric Fish

Among the many strange and beautiful fish in the seas, there is an e______ fish, capable of generating its own e______ with which it can give a severe shock.

The best known of these fish live in warm s______. They are called electric rays or torpedoes. T______ have an organ behind each eye which is made up of several hexagonal cells, like a honeycomb. These c______ contain a jelly-like substance and flat, electric plates.

The upper part of the o______ is the positive portion and the lower p______ is the negative portion. Very fine nerves connect the p______ and n______ sides of the cells, and these n______ carry the e______ to all parts of the t______ body.

This power of g______ electricity serves these fish for both defence and attack. When a______ by a larger fish, the torpedo will drive the attacker away by an e______ shock; but they can also use their electricity to sting smaller fish, so that they are easier to kill for food.

English Usage Test 5

1—Spelling and Punctuation

One **of the words in each of the following sentences has been mis-spelt. Write out each mis-spelt word correctly.**

1. How good is your memry?
2. A basic subject is arithematic.
3. Tomatos can be grown in back-gardens.
4. Are those your books over their?
5. The passingers were not allowed to smoke on the bus.
6. Gareth atempted to jump one and a half metres.
7. Sandra hated argueing with Sarah.
8. In Joe's garden were the most glorieus daffodils.
9. There are many beautiful islands off Greese.
10. Is your spelling akurate?

Punctuate **the following sentences:**

1. we went on a german ship to italy
2. help
3. are you going to help me asked gareth
4. the kittens were called smokey fatty tinker and fluff
5. the childrens old toys were sent to oxfam

English Usage Test 5

2—Vocabulary: synonyms

Write down the word in the second sentence which means the *same* as the word underlined in the first sentence.

1. The Mad Hatter was a friend of the White Rabbit.
   Was the man sane, insane or just silly?
2. Joe often goes training after school.
   Frequent rains caused a bad harvest.
3. The children created a game called Puggles.
   The statue was made by a famous sculptor.
4. Her skin was wrinkled at sixty-five.
   Winkled from its shell, the snail looked puckered.
5. Her smooth skin was without blemish.
   The blue diamond was spoiled by a flaw.

By writing *one* word in each of the spaces, make the second sentence mean the *same* as the first.

1. He would not give an answer.
   He ( ) to answer.
2. The audience was non-existent.
   There was ( ) there.
3. He was untroubled.
   He knew ( ) was ( ) ( ) worry ( ).
4. She was despised.
   She was ( ) down ( ).
5. The nations signed a non-aggression treaty.
   They signed a ( ) treaty.

English Usage Test 5

3—Vocabulary: antonyms

Write down the word in the second sentence that is *opposite* to the word underlined in the first sentence.

1. Gareth was going <u>to</u> Manchester.
   Two girls were coming from Newcastle.

2. Joe hid in a <u>hollow</u> tree.
   'Hello Joe' said the class swot. 'Do you know what a solid shape is?'

3. The army <u>advanced</u> to the battle-field.
   We departed for our holiday with three cases.

4. We rang for five minutes but there was no <u>reply</u>.
   'Answer my question, or we will have a replay' said the umpire.

5. I do not like very <u>sweet</u> chocolate.
   Sweat often has a sour smell.

If you were writing a story called 'Going to the Local Cinema', in what order would you use the following headings?

Going to the Local Cinema

Tall person sits in front of you.

Leaving cinema.

Moving to the front row before interval.

Getting a good seat in the back row.

Buying a choc-ice before main film.

English Usage Test 5

4—Sentence Construction

Form five different sentences by crossing out one or more of the words used in the sentence below. The first has been done for you.

You are a stupid and hurtful boy.
You are a ~~stupid and hurtful~~ boy.

Form the following groups of words into sentences by re-arranging them and adding *one* of your own.

1. cows horns bulls have
2. midnight clock the twelve
3. hairy Sarah frightened spiders was
4. Albert Gardens Kensington Memorial is the
5. Wellington Waterloo of the Battle

Do exactly the same for the next groups of words, only this time some of the words have been placed in order for you. Remember to add one word of your own.

6. cooks spoil broth many the
   ( ) many ( ) spoil ( ) ( ).
7. sheep comes wool
   ( ) comes ( ) ( ).
8. painter French was Matisse famous
   Matisse ( ) ( ) famous ( ) ( ).
9. strong roses most a fragrance
   Most ( ) ( ) a ( ) ( ).
10. creatures cats curious
    ( ) ( ) curious ( ).

English Usage Test 5

5—Reading and Comprehension

Read the following passage carefully, and then answer the questions below.

They reached the carriage-drive of Toad Hall to find, as the Badger had anticipated, a shiny new motor-car, of great size, painted a bright red (Toad's favourite colour), standing in front of the house. As they neared the door, it was flung open, and Mr. Toad, arrayed in goggles, cap, gaiters and enormous overcoat, came swaggering down the steps, drawing on his gauntleted gloves.

'Hullo! Come on, you fellows!' he cried cheerfully on catching sight of them. 'You're just in time to come with me for a jolly—to come for a jolly—for a—er—jolly—'

His hearty accents faltered and fell away, as he noticed the stern unbending look on the countenances of his silent friends, and his invitation remained unfinished.

The Badger strode up the steps. 'Take him inside' he said sternly to his companions. Then as Toad was hustled through the door, struggling and protesting, he turned to the chauffeur in charge of the new motor-car. 'I'm afraid you won't be wanted today' he said. 'Mr. Toad has changed his mind. He will not require the car. Please understand that this is final. You needn't wait.' Then he followed the others inside and shut the door.

'Now then,' he said to the Toad, when the four of them stood together in the hall, 'First of all, take these ridiculous things off!'

'Shan't!' replied Toad with great spirit. 'What is the meaning of this gross outrage? I demand an instant explanation.'

1. Describe Toad's feelings throughout the passage.
2. Why doesn't Toad finish his invitation?
3. Who ordered the chauffeur to put the car away?
4. Why were Toad's clothes ridiculous?
5. What do the following words mean: (a) anticipated
(b) faltered
(c) explanation?

English Usage Test 5

6—Reading and Comprehension

In the following passage, several words have been left out. The first letter of each missing word has been put in to help you. Write down the missing words, in the order in which they occur in the passage.

Movement and Music

If you stretch an elastic band between your index fingers and thumbs and pluck it, differently pitched noises can be produced.

A sweeter, or more m______ note can be made by vibrating tightly s______ string, wire or catgut. The different n______ are produced by altering the length of the wire.

The musical notes of the violin and s______ stringed instruments are made by shortening the l______ of the s______ with the fingers of one hand, while v______ them with a bow.

Inside a piano, you will see that the strings are tightened over an iron frame s______ to make the strings of different l______. All s______ instruments produce musical n______ from their vibrating strings.

When you whistle you are moving a______ to make musical notes. These are altered by shaping your l______ and tongue.

The third type of vibrating m______ which produces a musical s______ is banging. The drums and cymbals of an orchestra use this method to produce their particular sound.

English Usage Test 6

1—Spelling and Punctuation

***One* of the words in each of the following sentences has been mis-spelt. Write out each mis-spelt word correctly.**

1. The birthday tellegram came in a bright envelope.
2. Three lieutenants signed the treety.
3. Is nine o'clock a resonable bed-time?
4. Joe was dissappointed with the programme.
5. Sarah needed Gareth's asistance in crossing the river.
6. The seeling was painted navy-blue.
7. Sandra's spaniel always slept on a cowch.
8. It is vulger to poke out your tongue.
9. A storm-warning went out to all British shiping.
10. I cannot rekollect seeing you at our Wednesday-club.

Punctuate **the following sentences:**

1. s o s is a code signal of extreme distress
2. my mother asked me to go to a shop called belles
3. mr jones shop was painted pink
4. we ate a salad made of apples oranges and bananas
5. are you going to the match joe asked

English Usage Test 6

2—Vocabulary: synonyms

Write down the word in the second sentence which means the *same* as the word underlined in the first sentence.

1. There was a <u>lull</u> in the storm.
   The interval allowed us time to recover.

2. Sarah was <u>fatigued</u> by overwork at school.
   The fat policeman became tired fairly easily.

3. Robin Hood <u>camouflaged</u> himself by wearing green.
   The cauliflower was disguised in cheese sauce.

4. <u>Compare</u> the different polygons.
   Contrast this year's compère with the others.

5. Sandra was <u>disgruntled</u> at having eggs again.
   The pigs' grunts do not mean that they are discontented.

By writing *one* word in each of the spaces, make the second sentence mean the *same* as the first.

1. Rub the cat's fur one way only.
   (   ) the cat.

2. All the rooms opened on to one another.
   All the rooms were (   ).

3. The General rallied the troops.
   The troops were brought (   ) by the General.

4. Joe did not take advantage of his friends.
   Joe did not (   ) his friends.

5. She said she would not go.
   She (   ) to go.

English Usage Test 6

3—Vocabulary: antonyms

Write down the word in the second sentence that is *opposite* to the word underlined in the first sentence.

1. The tenant paid his rent promptly.
   Who is the owner of this tent?
2. Joe respected his father's advice.
   Despite his position, he was despised.
3. The ascent up the mountain was very difficult.
   The decent weather made the descent easy.
4. The French conquered the English in 1066.
   We were defeated and fled away.
5. The house was quite distant from the farm.
   There was a distinct smell of gas near the oven.

If you were writing a story called 'Making an Omelette', in what order would you use the following headings?

Making an Omelette

Fold omelette over.

Let mixture settle in frying-pan.

Tip out of pan, on to hot plate.

Whisk eggs, salt and pepper.

Break the eggs into a bowl.

English Usage Test 6

4—Sentence Construction

Form five different sentences by crossing out one or more of the words used in the sentence below. The first has been done for you.

The pigs squealed and grunted loudly.
The pigs squealed ~~and grunted~~ loudly.

Form the following groups of words into sentences by re-arranging them and adding *one* of your own.

1. one one hundred pound make
2. sweet flowers peas
3. unmarried an spinster is a
4. common salt seasonings pepper and
5. blood pumps your

Do exactly the same for the next groups of words, only this time some of the words have been placed in order for you. Remember to add one word of your own.

6. green make yellow blue
   Blue ( ) ( ) make ( ).
7. ice called frozen is
   ( ) ( ) is ( ) ice.
8. animal hide leather
   Leather ( ) ( ) ( ).
9. Paris is France capital the
   ( ) is ( ) capital ( ) ( ).
10. from sugar caramel is
    Caramel ( ) ( ) from ( ).

English Usage Test 6

5—Reading and Comprehension

Read the following passage carefully, and then answer the questions below.

The Uncles sat in front of the fire, took off their collars, loosened all buttons, put their large moist hands over their watch chains, groaned a little, and slept. Auntie Beattie, who liked port, stood in the middle of the snowbound back-yard, singing like a big-bosomed thrush. I would blow up balloons to see how big they blew up to; and when they burst, which they did, the Uncles jumped and rumbled. In the rich and heavy afternoon, the Uncles breathing like dolphins, and the snow descending, I would sit in the front room, among festoons and Chinese lanterns, and nibble at dates, and try to make a model man-o'-war, following the instructions for Little Engineers, and produce what might be mistaken for a sea-going tram. And then, at Christmas tea, the recovered Uncles would be jolly over their mince pies; and the great iced cake loomed in the centre of the table like a marble grave. Auntie Hannah laced her tea with rum, because it was only once a year. And in the evening there was music.

1. **Describe the Uncles.**
2. **How did the boy occupy himself in the afternoon?**
3. **What was Auntie Hannah like?**
4. **Was the boy successful in making his model ship? How do you know?**
5. **What do the following words mean: (a) moist**
   (b) descending
   (c) nibble?

English Usage Test 6

6—Reading and Comprehension

In the following passage, several words have been left out. The first letter of each missing word has been put in to help you. Write down the missing words, in the order in which they occur in the passage.

Electricity

A long time ago, it was discovered that if a piece of amber was rubbed on a dry cloth, it attracted tiny pieces of paper. This attracting force was called e______, taking its name from the Greek word elektron, meaning a______.

The first simple dynamo was invented by Michael Faraday, who discovered that electric currents could be produced by moving coils of wire in the field of a magnet.

A d______ is a machine that spins a c______ of w______ between the poles of a m______. This action causes an electric c______ to pass through the w______. The faster the coil of wire is s______ the more electricity is generated.

In large power stations, steam turbines are used to spin the coils of wires at tremendous s______ between the p______ of magnets, and the electric c______ g______ is carried by cables for use in your homes.

Sometimes dams are built, not just as reservoirs, but so that the water can be piped to a water t______ which drives the d______.

Mathematics Tests: Answers

Test 1

1—Basic Computation

1. e
2. a
3. f
4. c
5. g
6. 3
7. 14
8. 4
9. 3, 9 or 9, 3
10. $\frac{3}{10}$
11. £2.97
12. 11
13. 68p
14. 63p
15. 48p

2—Understanding of Number

1. g
2. c
3. f
4. h
5. b
6. 5568
7. $\frac{4}{9}$
8. 3 ÷ 16
9. c
10. b
11. 3287
12. 2238

3—Understanding of Number

	Large Squares	Rows	Left Over
13.	1	1	2
14.	2	1	1
15.	2	0	0
16.	1	0	1
17.	2	2	2

18. 16

4—Sets, Sequences and Reasoning

1. d
2. c
3. 2
4. 3
5. d

5—Sets, Sequences and Reasoning

6. [23, 27, 31, 35] 39, 43
7. [29, 37, 45, 53] 61, 69
8. [25, 37, 49, 61] 73, 85
9. [2, 6, 18, 54] 162, 486
10. [3, 6, 12, 24] 48, 96
11. [7, 8, 10, 13] 17, 22
12. [5, 7, 11, 17] 25, 35
13. [$\frac{1}{1}$, $\frac{1}{3}$, $\frac{1}{9}$, $\frac{1}{27}$] $\frac{1}{81}$, $\frac{1}{243}$
14. [$\frac{2}{1}$, $\frac{2}{2}$, $\frac{2}{4}$, $\frac{2}{8}$] $\frac{2}{16}$, $\frac{2}{32}$
15. [$\frac{2}{4}$, $\frac{6}{8}$, $\frac{10}{12}$, $\frac{14}{16}$] $\frac{18}{20}$, $\frac{22}{24}$
16. 7
17. 5

6—Spatial Discrimination

1. yes—cuboid
2. no
3. yes—tetrahedron
4. yes—triangular prism
5. yes—hexagonal prism
6. no
7. yes
8. no
9. yes
10. yes
11. yes
12. no

7—Measurements and Measuring

1. Any number between 600 and 720
2. Any number between 8 and 50
3. Any number between 500 and 700
4. Any number greater than 2000
5. Any number less than 125
6. 33 hours 5 minutes
7. 15 hours 25 minutes
8. Roughly 60 kilometres
9. Roughly 60 kilometres
10. Roughly 98 kilometres

8—Measurements and Measuring

11. Roughly 114
12. Roughly 210 seconds
13. Roughly 67 seconds
14. Roughly 25 seconds
15. Roughly 215 seconds
16. Roughly 85 times

Test 2

1—Basic Computation

1. g
2. a
3. f
4. e
5. d
6. 6
7. 14
8. 8
9. 9, 5 or 5, 9 or 3, 15 or 15, 3
10. $\frac{1}{4}$
11. £6.32
12. 5
13. 53p
14. 72p
15. 65p

2—Understanding of Number

1. e
2. g
3. h
4. d
5. f
6. 3528
7. $\frac{13}{32}$
8. 0.3
9. d
10. b
11. 7352
12. 1278

3—Understanding of Number

	Large Squares	Rows	Left Over
13.	2	2	3
14.	1	2	1
15.	1	0	3
16.	1	0	0
17.	2	0	2

18. 49
19. 8

4—Sets, Sequences and Reasoning

1. e
2. b
3. 4
4. 7
5. c

5—Sets, Sequences and Reasoning

6. [49, 51, 53, 55] 57, 59
7. [28, 35, 42, 49] 56, 63
8. [10, 23, 36, 49] 62, 75
9. [5, 10, 20, 40] 80, 160
10. [7, 21, 63, 189] 567, 1701
11. [8, 9, 11, 15] 23, 39
12. [10, 11, 14, 19] 26, 35
13. $[\frac{1}{2}, \frac{1}{1}, \frac{3}{2}, \frac{2}{1}]\ \frac{5}{2}, \frac{3}{1}$
14. $[\frac{1}{8}, \frac{1}{4}, \frac{3}{8}, \frac{1}{2}]\ \frac{5}{8}, \frac{3}{4}$
15. $[\frac{1}{3}, \frac{1}{6}, \frac{1}{12}, \frac{1}{24}]\ \frac{1}{48}, \frac{1}{96}$
16. 21
17. 17

6—Spatial Discrimination

1. yes—cuboid
2. no
3. no
4. yes—square pyramid
5. yes—cube
6. yes—truncated pyramid
7. yes
8. yes
9. no
10. yes
11. yes
12. no

7—Measurements and Measuring

1. Any number between 6 and 30
2. Any number between 30 and 100
3. Any number between 40 and 100
4. Any number between 200 and 500
5. Any number less than 250 000
6. 31 hours 20 minutes
7. 29 hours 10 minutes
8. Roughly 140 kilometres
9. Roughly 360 kilometres
10. Roughly 176 kilometres

8—Measurements and Measuring

11. Roughly 4.5 km/hr
12. 4 minutes
13. Roughly 1 min 45 secs
14. Roughly 50 seconds
15. Roughly 5 minutes
16. Roughly 200 metres

Test 3

1—Basic Computation

1. c
2. e
3. g
4. f
5. b
6. 22
7. 18
8. 7
9. 4, 6 or 6, 4 or 2, 12 or 12, 2 or 8, 3 or 3, 8
10. $\frac{5}{12}$
11. £3.67
12. 29p
13. £1.05
14. 32p
15. 18p

2—Understanding of Number

1. g
2. h
3. d
4. e
5. b
6. 7678
7. $\frac{3}{10}$
8. $\frac{1}{8}$
9. c
10. b
11. 4527
12. 23 578

3—Understanding of Number

	Large Squares	Rows	Left Over
13.	1	1	5
14.	1	0	2
15.	1	1	0
16.	1	3	2
17.	2	0	1

18. 25
19. 10

4—Sets, Sequences and Reasoning

1. d
2. d
3. 6
4. 1
5. b

5—Sets, Sequences and Reasoning

6. [57, 59, 61, 63] 65, 67
7. [66, 71, 76, 81] 86, 91
8. [31, 47, 63, 79] 95, 111
9. [2, 8, 32, 128] 512, 2048
10. [4, 8, 16, 32] 64, 128
11. [2, 2, 5, 11] 20, 32
12. [1, 3, 7, 15] 31, 63
13. $[2, \frac{9}{4}, \frac{5}{2}, \frac{11}{4}]\ 3, \frac{13}{4}$
14. $[\frac{1}{4}, \frac{5}{16}, \frac{3}{8}, \frac{7}{16}]\ \frac{1}{2}, \frac{9}{16}$
15. $[\frac{1}{2}, \frac{17}{32}, \frac{9}{16}, \frac{19}{32}]\ \frac{5}{8}, \frac{21}{32}$
16. 8
17. 15

6—Spatial Discrimination

1. yes—triangular prism
2. yes—octahedron
3. no
4. no
5. yes—prism
6. yes—hexagonal pyramid
7. yes
8. no
9. yes
10. no
11. yes
12. yes

7—Measurements and Measuring

1. Any number between 420 and 480
2. Any number between 3 and 5
3. Any number between 800 and 1000
4. Any number between 2 and 5
5. Any number less than 100 000
6. 13 hours 40 minutes
7. 41 hours 35 minutes
8. Roughly 575 kilometres
9. Roughly 275 kilometres
10. Roughly 150 kilometres

8—Measurements and Measuring

11. About 52 paces per min
12. 10
13. About 20 secs
14. About 60 secs
15. About 3 mins
16. After about $6\frac{1}{2}$ mins

Test 4

1—Basic Computation

1. h
2. d
3. e
4. a
5. c
6. 27
7. 16
8. 7
9. 2, 7 or 7, 2
10. $\frac{1}{6}$
11. £5.51
12. 18
13. 78p
14. 80p
15. 18p

2—Understanding of Number

1. e
2. g
3. b
4. f
5. h
6. 6678
7. $\frac{7}{8}$
8. 23÷12
9. b
10. c
11. 5068
12. 8443

3—Understanding of Number

	Large Cubes	Squares	Rows	Left Over
13.	1	2	0	1
14.	1	1	0	2
15.	1	2	2	2
16.	2	1	1	0
17.	2	2	0	0

18. 64

4—Sets, Sequences and Reasoning

1. c
2. b
3. □ large
4. ○ small
5. b

5—Sets, Sequences and Reasoning

6. [38, 41, 44, 47] 50, 53
7. [45, 54, 63, 72] 81, 90
8. [21, 36, 51, 66] 81, 96
9. [7, 14, 28, 56] 112, 224
10. [1, 3, 7, 9, 13] 15, 19
11. [1, 4, 10, 13, 19] 22, 28
12. [3, 5, 9, 15] 23, 39
13. [4, 7, 13, 22] 34, 49
14. [$\frac{2}{3}$, 1, $1\frac{2}{3}$, $2\frac{2}{3}$] 4, $5\frac{2}{3}$
15. [$\frac{3}{8}$, $\frac{1}{2}$, $\frac{3}{4}$, $\frac{9}{8}$] $\frac{13}{8}$, $\frac{9}{4}$
16. 6
17. 12

6—Spatial Discrimination

1. yes—triangular prism
2. yes—triangular prism
3. no
4. yes—hexagonal prism
5. yes—pentagonal pyramid
6. no
7. yes
8. no
9. yes
10. no
11. yes
12. yes

7—Measurements and Measuring

1. Any number between 11 and 15
2. 9
3. Any number between 3 and 10
4. Any number less than 20
5. Any number less than 50
6. 73 hours 35 minutes
7. 71 hours 30 minutes
8. Roughly 1200 kilometres
9. Roughly 1600 kilometres
10. Roughly 900 kilometres

8—Measurements and Measuring

11. 55 kilometres per hour
12. 8
13. Roughly 110 secs
14. Roughly 80 secs
15. Roughly 140 secs
16. Roughly $\frac{5}{6}$ of a kilometre

Test 5

1—Basic Computation

1. f
2. a
3. d
4. c
5. h
6. 33
7. 3
8. 4
9. 5, 5
10. $\frac{1}{16}$
11. £3.88
12. 19
13. 17p
14. 72p
15. 70p

2—Understanding of Number

1. g
2. b
3. h
4. d
5. c
6. 2238
7. $\frac{7}{3}$
8. 17÷6
9. a
10. b
11. 18 430
12. 9972

3—Understanding of Number

	Large Cubes	Squares	Rows	Left Over
13.		2	1	0
14.		2	3	1
15.		3	2	0
16.	1	0	2	3
17.	1	2	0	1

18. 125
19. 3

4—Sets, Sequences and Reasoning

1. b
2. d
3. ○ large
4. ● small
5. a

5—Sets, Sequences and Reasoning

6. [52, 56, 60, 64] 68, 72
7. [31, 38, 45, 52] 59, 66
8. [20, 34, 48, 62] 76, 90
9. [8, 24, 72, 216] 648, 1944

10. [5, 7, 10, 12, 15] 17, 20
11. [7, 10, 15, 18, 23] 26, 31
12. [3, 3, 6, 18] 72, 360
13. [5, 10, 40, 240] 1920, 19 200
14. [$\frac{2}{5}$, $\frac{3}{5}$, 1, $\frac{8}{5}$] $\frac{12}{5}$, $\frac{17}{5}$
15. [$\frac{5}{6}$, $\frac{5}{9}$, $\frac{10}{27}$, $\frac{20}{81}$] $\frac{40}{243}$, $\frac{80}{729}$
16. 9
17. 20

6—Spatial Discrimination

1. no
2. yes—square pyramid
3. yes—prism
4. yes—triangular pyramid
5. no
6. no
7. yes
8. yes
9. no
10. no
11. yes
12. yes

7—Measurements and Measuring

1. Any number between 180 and 240
2. Any number between 900 and 1000
3. Any number between 20 and 100
4. Any number between 800 and 900
5. Any number less than 20
6. 35 hours 15 minutes
7. 35 hours 45 minutes
8. Roughly 4000 kilometres
9. Roughly 6750 kilometres
10. Roughly 2750 kilometres

8—Measurements and Measuring

11. Roughly 38 pedal turns per minute
12. After about 8 minutes 40 seconds
13. About 35 seconds
14. About 1 minute
15. About 145 seconds
16. About 32 times

Test 6

1—Basic Computation

1. h
2. d
3. c
4. a
5. e
6. 26
7. 25
8. 5
9. 5, 3 or 3, 5
10. $\frac{7}{8}$
11. £3.80
12. 16
13. 19p
14. 48p
15. 94p

2—Understanding of Number

1. h
2. e
3. g
4. b
5. f
6. 3258
7. $\frac{17}{16}$
8. $2\frac{1}{2}$
9. b
10. a
11. 19 308
12. 87 533

3—Understanding of Number

	Large Cubes	Squares	Rows	Left Over
13.		1	5	5
14.		2	1	5
15.		1	3	4
16.		2	3	3
17.	1	1	1	4
18.	216			
19.	8			

4—Sets, Sequences and Reasoning

1. c
2. b
3. ■ large
4. ○ large
5. d

5—Sets, Sequences and Reasoning

6. [76, 81, 86, 91] 96, 101
7. [13, 21, 29, 37] 45, 53
8. [27, 43, 59, 75] 91, 107
9. [4, 12, 36, 108] 324, 972
10. [3, 6, 10, 13, 17] 20, 24
11. [4, 9, 12, 17, 20] 25, 28
12. [5, 5, 10, 30] 120, 600
13. [1, 1, 3, 15] 105, 945
14. [$\frac{4}{7}$, $\frac{5}{7}$, 1, $\frac{10}{7}$] 2, $\frac{19}{7}$
15. [1, $\frac{3}{5}$, $\frac{9}{25}$, $\frac{27}{125}$] $\frac{81}{625}$, $\frac{243}{3125}$
16. 14
17. 18

6—Spatial Discrimination

1. yes—hexagonal pyramid
2. no
3. yes—truncated pyramid
4. no
5. yes—triangular pyramid
6. yes—prism
7. no
8. yes
9. yes
10. yes
11. no
12. yes

7—Measurements and Measuring

1. Any number between 11 and 30
2. Any number between 90 and 100
3. Any number between 40 and 50
4. Any number between 6 and 10
5. Any number greater than 1 000 000
6. 59 hours 25 minutes
7. 35 hours 35 minutes
8. Roughly 500 kilometres
9. Roughly 400 kilometres
10. Roughly 270 kilometres

8—Measurements and Measuring

11. About 35 breaths per min
12. 5
13. About 90 secs
14. About 150 secs
15. About 330 secs
16. About 64

English Usage Tests: Answers

Test 1

1—Spelling and Punctuation

1. really
2. history
3. liquid
4. music
5. medium
6. mathematics
7. omelet(te)
8. mirrors
9. exhausted
10. carpet

1. The river Thames flows through London.
2. For supper we cooked sausages, chips and beans.
3. 'Is there enough food for breakfast?' asked Sarah.
4. Don't do that.
5. All Joe's trousers were Levis.

2—Vocabulary: synonyms

1. terrible
2. ill
3. asked
4. fashionable
5. moaned

1. In school we must do games.
2. They were happy/pleased to have won the game.
3. Gareth remembered the birthday card.
4. Carrying the vase made her hands shake.
5. I like your cooking very much.

3—Vocabulary: antonyms

1. submerged
2. exposed
3. sold
4. hate
5. failure

A Night at Green Mansions

Going up to bed.
Waking and seeing a hand on the pillow.
The hand disappears and a skull appears.
See ghostly hand reappear.
Faint with fright.

4—Sentence Construction

The astronauts signalled messages.
The astronauts signalled.
The astronauts received messages.
The astronauts received and signalled.
Astronauts signalled messages.

1. Gareth played in the first cricket eleven.
2. Joe scored the first goal in the football match.
3. It was the best birthday party Sarah had ever had.
4. The referee wore knee-length shorts.
5. Sandra hated to play the game of snap.
6. (Last) (Sunday) (it) rained (all) day.
7. The (toffee) (caused) Gareth's (filling) (to) fall (out).
8. (The) children (went) to (the) (country) for (Easter).
9. I did (not) (know) whether (she) was (laughing) or (crying).
10. Blondes (have) (fair) hair.

6—Reading and Comprehension

1. camouflage
2. enemy
3. unseen
4. coloured
5. resembling
6. habitat
7. inhabit
8. attach
9. colour
10. creatures
11. moves
12. reddish (reddy)
13. verge
14. once
15. camouflage

Test 2

1—Spelling and Punctuation

1. pillar
2. grocer
3. cellar
4. chimneys
5. mention
6. parrots
7. handkerchiefs
8. capital
9. headache
10. vegetable

1. Sarah's teeth were the whitest in the class.
2. How much pocket money do you get?
3. 'Please be quiet,' said Miss Roberts.
4. We put the paints, brushes, books and crayons in the trunk.
5. Name the country nearest to New Zealand.

2—Vocabulary: synonyms

1. convinced
2. pursuit
3. link
4. pupils
5. deduct

1. While she was away the flat was burgled.
2. He pretended not to have heard.
3. I hate jellybabies.
4. The clock struck midnight.
5. Sandra was going my way.

3—Vocabulary: antonyms

1. old
2. whole
3. detached
4. front
5. mean

A Day's Fishing

Digging for very fat, juicy worms.
Baiting the line.
Float pulled under water.
Enormous fish escapes off hook.
Telling friend about the one which got away.

4—Sentence Construction

When will she do it?
Will she have time to do it?
Will she have time?
When will she have time?
Will she do it?

1. Polish is a foreign language.
2. Why is gold valuable?
3. What is the twelfth letter of the alphabet?
4. There are sixty minutes in an hour.
5. We went trout fishing.
6. Joe (was) (born) (under) the sign (of) (Scorpio).
7. Do (you) (believe) (in) horoscopes?
8. (Hades) was the (Greek) (god) of (the) (underworld).
9. (The) chocolate (was) (filled) (with) coffee.
10. She (promised) (to) marry (him).

6—Reading and Comprehension

1. bees
2. workers
3. community
4. honey
5. nectar
6. hairy
7. collecting
8. sac
9. honey
10. hungry
11. food
12. flowers
13. delivers
14. back
15. hive

Test 3

1—Spelling and Punctuation

1. suggest
2. ancient
3. peculiar
4. mysterious
5. generally
6. hymns
7. finally
8. guards
9. through
10. gnawed

1. Shall we buy him a toy, sweets, or a book?
2. 'May I borrow your gun?' asked Joe.
3. Sarah supported Manchester United.
4. She's got blonde hair.
5. Gareth's hair was brown.

2—Vocabulary: synonyms

1. swopped
2. treaty
3. doubtful
4. famous
5. glossy

1. During my absence they changed the team.
2. Keep away from railway lines.
3. He knew that his team had been relegated.
4. He baited his hook.
5. She replaced the books.

3—Vocabulary: antonyms

1. serious
2. accepted
3. screamed
4. advanced
5. given

Holiday in a Tent

Finding a good site.
Pitching the tent.
Getting rained out.
Slithering into wet sleeping bags.
Back to sleep in van.

4—Sentence Construction

Shout her name.
Don't shout.
Don't shout out.
Shout out her name.
Don't shout her name.

1. Will the door have a bolt or a key?
2. Milk is very good for you.
3. Denim jeans are very hard wearing.
4. Cheese is made from milk.
5. 'Wind in the Willows' was written by Kenneth Grahame.
6. The statue (of) (Eros) is (in) (Piccadilly) (Circus).
7. Wild (violets) (are) sometimes (white) in (colour).
8. A (cavalry) (regiment) rides (horses).
9. (During) an eclipse (the) moon (covers) (the) (sun).
10. (The) Italian goddess (of) flowers (was) (called) (Flora).

6—Reading and Comprehension

1. Golden
2. peninsulas
3. bridge
4. peninsula
5. Bay
6. water
7. pier
8. shore
9. anchored
10. cable
11. strands
12. lane
13. single
14. spans
15. overall

Test 4

1—Spelling and Punctuation

1. carriages
2. brief
3. jealous
4. muscles
5. circus
6. recently
7. similar
8. piece
9. orange
10. eagle

1. Who is your favourite singer?
2. Sarah's knee was badly bruised.
3. 'He's got lots of freckles,' said Sandra.
4. She whispered across to Joe, 'Do you want Sarah to come?'
5. You will marry a tinker, tailor, soldier or sailor.

2—Vocabulary: synonyms

1. bottom
2. combined
3. slyness
4. listened
5. refused

1. You resemble your father.
2. Our leader ordered us to put out the fire.
3. She found it impossible to lie.
4. The actress used / exploited her beauty.
5. She admired her brother.

3—Vocabulary: antonyms

1. calm
2. native
3. part
4. common
5. unfair

School Play

Auditioning for parts in school play.
Rehearsals after school.
Final dress rehearsal.
Curtain up! First night.
Applause and shouts for more.

4—Sentence Construction

Hector and the Trojans planned.
Hector planned.
The Trojans planned.
The Trojans planned the attack.
Hector planned the attack.

1. I like watching television on Saturdays.
2. Helen of Troy was famous for her beauty.
3. Copper Beech trees have very red leaves.
4. Biting your nails is a habit.
5. Do you carry your books in a satchel?
6. Every (fourth) (year) has (an) (extra) (day).
7. (Sunflower) seeds (are) good (for) (parrots).
8. (Gareth) (and) Joe built (a) (log) (cabin) (by) the (river).
9. (Sarah) made (few) (mistakes) (in) mathematics.
10. (Tigers) have (white) (spots) on (their) (ears).

6—Reading and Comprehension

1. electric
2. electricity
3. seas
4. Torpedoes
5. cells
6. organ
7. part
8. positive
9. negative
10. nerves
11. electricity
12. torpedo's
13. generating
14. attacked
15. electric

Test 5

1—Spelling and Punctuation

1. memory
2. arithmetic
3. tomatoes
4. there
5. passengers
6. attempted
7. arguing
8. glorious
9. Greece
10. accurate

1. We went on a German ship to Italy.
2. Help!
3. 'Are you going to help me?' asked Gareth.
4. The kittens were called Smokey, Fatty, Tinker and Fluff.
5. The children's old toys were sent to Oxfam.

2—Vocabulary: synonyms

1. insane
2. frequent
3. made
4. puckered
5. flaw

1. He refused to answer.
2. There was nobody there.
3. He knew there was nothing to worry about.
4. She was looked down on.
5. They signed a peace treaty.

3—Vocabulary: antonyms

1. from
2. solid
3. departed
4. question
5. sour

Going to the Local Cinema

Getting a good seat in the back row.
Tall person sits in front of you.
Moving to the front row before interval.
Buying a choc-ice before main film.
Leaving cinema.

4—Sentence Construction

You are a boy.
You are a stupid boy.
You are a hurtful boy.
You are stupid.
You are hurtful.

1. Cows and bulls have horns.
2. The clock struck twelve midnight.
3. Sarah was frightened of hairy spiders.
4. The Albert Memorial is in Kensington Gardens.
5. Wellington won the Battle of Waterloo.
6. (Too) many (cooks) spoil (the) (broth).
7. (Wool) comes (from) (sheep).
8. Matisse (was) (a) famous (French) (painter).
9. Most (roses) (have) a (strong) (fragrance).
10. (Cats) (are) curious (creatures).

6—Reading and Comprehension

1. musical	6. string	11. notes
2. stretched	7. vibrating	12. air
3. notes	8. shaped	13. lips
4. similar	9. lengths	14. movement
5. length	10. stringed	15. sound

Test 6

1—Spelling and Punctuation

1. telegram	5. assistance	9. shipping
2. treaty	6. ceiling	10. recollect
3. reasonable	7. couch	
4. disappointed	8. vulgar	

1. S.O.S. is a code signal of extreme distress.
2. My mother asked me to go to a shop called Belles.
3. Mr. Jones' shop was painted pink.
4. We ate a salad made of apples, oranges and bananas.
5. 'Are you going to the match?' Joe asked.

2—Vocabulary: synonyms

1. interval	4. contrast
2. tired	5. discontented
3. disguised	

1. Stroke the cat.
2. All the rooms were connecting/connected.
3. The troops were brought together by the General.
4. Joe did not use his friends.
5. She refused to go.

3—Vocabulary: antonyms

1. owners	4. defeated
2. despised	5. near
3. descent	

Making an Omelette

Break the eggs into bowl.
Whisk eggs, salt and pepper.
Let mixture settle in frying-pan.
Fold omelette over.
Tip out of pan on to hot plate.

4—Sentence Construction

The pigs squealed loudly.
The pigs squealed.
The pigs grunted.
The pigs grunted loudly.
The pigs squealed and grunted.

1. One hundred pence make one pound.
2. Sweet peas are flowers.
3. A spinster is an unmarried woman.
4. Salt and pepper are common seasonings.
5. Your heart pumps blood.
6. Blue (and) (yellow) make (green).
7. (Frozen) (water) is (called) ice.
8. Leather (is) (animal) (hide).
9. (Paris) is (the) capital (of) (France).
10. Caramel (is) (made) from (sugar).

6—Reading and Comprehension

1. electricity	6. magnet	11. poles
2. amber	7. current	12. currents
3. dynamo	8. wire	13. generated
4. coil	9. spun	14. turbine
5. wire	10. speed	15. dynamo